RURAL BUILDING

VOLUME FOUR

Drawing Books

INTERMEDIATE TECHNOLOGY PUBLICATIONS 1995

Published by Intermediate Technology Publications Ltd,
103–105 Southampton Row, London WC1B 4HH, UK
Under licence from TOOL, Sarphatistraat 650, 1018 AV, Amsterdam, Holland

© TOOL

A CIP catalogue record for this book
is available from the British Library

ISBN 1 85339 325 8

Printed in Great Britain by SRP, Exeter, UK

P R E F A C E

This official text book is designed purposely to meet the needs of trainees who are pursuing rural building courses in various training centres administered by the National Vocational Training Institute.

The main aim of this book is to provide much needed trade information in simple language and with illustrations suited to the understanding of the average trainee.

It is the outcome of many years of experiment conducted by the Catholic F.I.C. brothers of the Netherlands, and the German Volunteer Service instructors, in simple building techniques required for a rural community.

The National Vocational Training Institute is very grateful to Brothers John v. Winden and Marcel de Keijzer of F.I.C. and Messrs. Fritz Hohnerlein and Wolfram Pforte for their devoted service in preparing the necessary materials for the book; we are also grateful to the German Volunteer Service and the German Foundation For International Development (DSE) - AUT, who sponsored the publication of this book.

We are confident that the book will be of immense value to the instructors and trainees in our training centres.

DIRECTOR: National Vocational Training Institute, Accra

© Copyright: by Stichting Kongregatie F.I.C. Brusselsestraat 38 - 6211 PG Maastricht/Nederland
Alle Rechte vorbehalten - All rights reserved.

INTRODUCTION TO A RURAL BUILDING COURSE

Vocational training in Rural Building started in the Nandom Practical Vocational Centre in 1970. Since then this training has developed into an official four year course with a programme emphasis on realistic vocational training.

At the end of 1972 the Rural Building Course was officially recognised by the National Vocational Training Institute. This institute guides and controls all the vocational training in Ghana, supervises the development of crafts, and sets the examinations that are taken at the end of the training periods.

The Rural Building programme combines carpentry and masonry, especially the techniques required for constructing housing and building sanitary and washing facilities, and storage facilities. The course is adapted to suit conditions in the rural areas and will be useful to those interested in rural development, and to farmers and agricultural workers.

While following this course, the instructor should try to foster in the trainee a sense of pride in his traditional way of building and design which is influenced by customs, climate and belief. The trainee should be aware of the requirements of modern society, and the links between old and new techniques, between modern and traditional designs -- and how best to strike a happy medium between the two with regard to considerations like health protection, storage space, sewage, and the water supply. The trainee should be encouraged to judge situations in the light of his own knowledge gained from the course, and to find his own solutions to problems; that is why this course does not provide fixed solutions but rather gives basic technical information. The instructor can adapt the course to the particular situation with which he and the trainee are faced.

This course is the result of many years of work and experimentation with different techniques. The text has been frequently revised to serve all those interested in Rural Development, and it is hoped that this course will be used in many vocational centres and communities. It is also the sincere wish of the founders of this course that the trainees should feel at the completion of their training that they are able to contribute personally to the development of the rural areas, which is of such vital importance to any other general development.

We are grateful to the Brothers F.I.C., the National Vocational Training Institute, and the German Volunteer Service for their support and assistance during the preparation of this course.

Bro. John v. Winden (F.I.C.)

LAY-OUT OF THE RURAL BUILDING COURSE

The Rural Building Course is a block-release-system course, which means that the trainee will be trained in turn at the vocational centre and at the building site. The period of training at the centre is called "off-the-job training", and the period on the building site is called "on-the-job training". Each will last for two years, so that the whole course will take four years and will end with the final test for the National Craftsmanship Certificate.

BLOCK RELEASE SYSTEM

YEAR	TERM 1	TERM 2	TERM 3
1	X	X	X
2	O	O	O
3	O	X	O
4	X	O	X

X = OFF-THE-JOB TRAINING
O = ON-THE-JOB TRAINING

The total "off-the-job" training period lasts approximately 76 weeks, each week 35 hours. During this training about 80% of the time is spent on practical training in the workshop. The remaining 20% of the time is devoted to theoretical instruction.

The total "on-the-job" training period lasts approximately 95 weeks, each week 40 hours. During this period the trainee does full-time practical work related to his course work. In addition some "homework" is assigned by the centre and checked by the instructors.

A set of books has been prepared as an aid to the theoretical training:
    Rural Building, Basic Knowledge (Form 1)
    Rural Building, Construction (Forms 2, 3, 4)
    Rural Building, Drawing Book (Forms 1, 2, 3, 4)
    Rural Building, Reference Book

All these books are related to each other and should be used together. The whole set covers the syllabus for Rural Building and will be used in the preparation for the Grade II, Grade I, and the National Craftsmanship Certificate in Rural Building.

SYLLABUS FOR RURAL BUILDING DRAWING

FORM I

Drawing equipment (pages 1-2)
Lines and lettering (pages 3-5)
Orthographic drawing (pages 6-9)
Oblique drawing (pages 10-13)
Oblique drawing to orthographic drawing and orthographic to oblique (pages 14-21)
Scale drawing (pages 22-23)
Inside and outside dimensions (pages 24-25)
Designing from sketches (pages 26-31)
Oblique and orthographic drawings of a box-like object (pages 32-33)
Cross sections (pages 34-35)
Sketching (pages 36-39)
Window frames (pages 40-42)
Door frames (page 43)
Frames and joints (pages 44-47)

FORM II

General building information (page 48)
Building drawing key (pages 49-50)
How buildings are drawn (pages 51-54)
Floor plans (pages 55-58)
Elevations (pages 59-60)
Sections and cross sections (pages 61-62)
Plans - elevations - cross sections (page 63)
Foundation plans (pages 64-66)
Door frames (pages 67-71)
Window frames (page 72)
Casements and doors (pages 73-78)

FORM III

Working drawings (page 79)

Building with pentroof (pages 81–82)

Pentroof plan (page 83)

Parapetted pentroof (page 84)

Parapetted gable roof (pages 85–86)

Building with gable roof (pages 87–88)

Gable roof plan (page 89)

Building with verandah (pages 90–92)

Foundation plan (page 93)

Gable roof with overhang design (page 94)

FORM IV

The Rural Builder (page 96)

Building design (pages 97–98)

Location plan (pages 99–103)

Boundary line (page 101)

Building information (pages 104–106)

Basic outline of different buildings and roofs (pages 107–116)

Circular work (page 117)

Grain silo (page 118)

Water well (page 119)

Water filter (page 120)

Pit latrine and aqua privy (page 121)

Pit latrine and squatting slab (page 122)

Bucket latrine (page 123)

Community pit latrine (page 124)

Manhole (page 125)

Septic tank (page 126)

BOOK INTRODUCTION

The drawing book is divided into four sections, corresponding to the four forms in the Rural Building curriculum.

The lessons are planned to last approximately 90 minutes, during which the instructor should spend some time going over mistakes made in the drawings from previous lessons. Then the instructor can give the introduction to the new lesson, following this by a discussion of the how and why of the lesson. At the end of the session, the instructor should furnish the trainees with the technical data for the new drawing they are assigned for that week. Assignments should be handed in a few days before the next lesson so the instructor has time to correct them and make himself acquainted with the general difficulties which appear as a result.

Tests should be given at intervals. The drawings should then be made within the time specified in the book. The instructor can write the necessary technical data on the blackboard.

In the first part of the book much emphasis is put on the techniques for oblique and orthographic drawings. This kind of drawing has to be mastered early for the trainee to be able to understand drawings made on the blackboard during lessons. Sketching is important because during practicals many explanations are made with the aid of sketches. Plenty of time should be allowed for these exercises, and it is only at the end of the first year that the trainee should attempt drawings of simple frames. The instructor can add other drawings or sketches as needed to help the trainees to understand.

The last three parts of the book are oriented towards the course content for Rural Building. In these parts the trainee will find the lay-out of a whole building from the foundations to the roof construction, with the plans, elevations and cross sections; and building design is discussed as well. Here too, the instructor should feel free to change the sequence of the lessons if necessary to fit them together with the practicals in the workshop. During the lessons the instructor is advised to visit building sites with the trainees so that they can compare the drawings with the actual structures.

Ability to read drawings is also very important and ample time should be spent to help the trainees master this. It is helpful at times to have the trainees exchange drawings and correct each other's work.

DRAWING EQUIPMENT

DRAWING BOARD

A drawing board should be made from well seasoned wood or good quality plywood. One edge of the board -- usually the left edge -- should be very straight so it can be used with the T-square.

T-SQUARE

T-squares are rulers with a cross piece or stock fixed on one end to form the letter "T". The stock is either glued or attached to the blade by screws. Like drawing boards, T-squares should be made from materials which do not warp easily.

SET SQUARES

Set squares are triangular shaped tools which are used with the T-square. Two angles are available; 45 degrees and 60 degrees.

RULERS

Rulers used for drawing have 30 centimetre (cm) scales, subdivided into 300 millimetres (mm).

PENCILS

Pencils used for drawing are usually 2H, 3H, or 4H. The higher the number, the harder the lead. Sharpen pencils with a pocket knife. Cut the wood at a low angle to expose about 7 mm of lead, then sharpen the lead by carefully rubbing it on a piece of sandpaper.

Pencil lines must be fine, light and clear. It is a good habit to rotate the pencil as you draw a line, to keep a sharp point on the lead. When drawing lines follow the instructions on page 3.

DRAWING PINS

Drawing pins are used to fix paper to the board. The pins should have short fine points so that they don't make large holes in the drawing board. The pins should have large flat heads so that they can be removed easily. Because drawing pins can damage the T-square, various types of adhesive (sticky) tapes are often used instead of pins.

THE COMPASS

The compass is a precision instrument used to draw circles. One of the legs has a pointed end; this point must be thin and sharp so that it makes only a small hole in the paper. Especially when drawing small circles, make sure that the pencil point is the same length as the steel pin.

ERASERS

An eraser should be soft and of good quality so it does not damage the paper. The eraser should be used very little, and only with great care. If the corners of the eraser are rounded, it is a good idea to cut one end sharp again if you need to erase very exactly.

DRAWING PAPER

Drawing paper is special paper and it is cut to standard sizes.

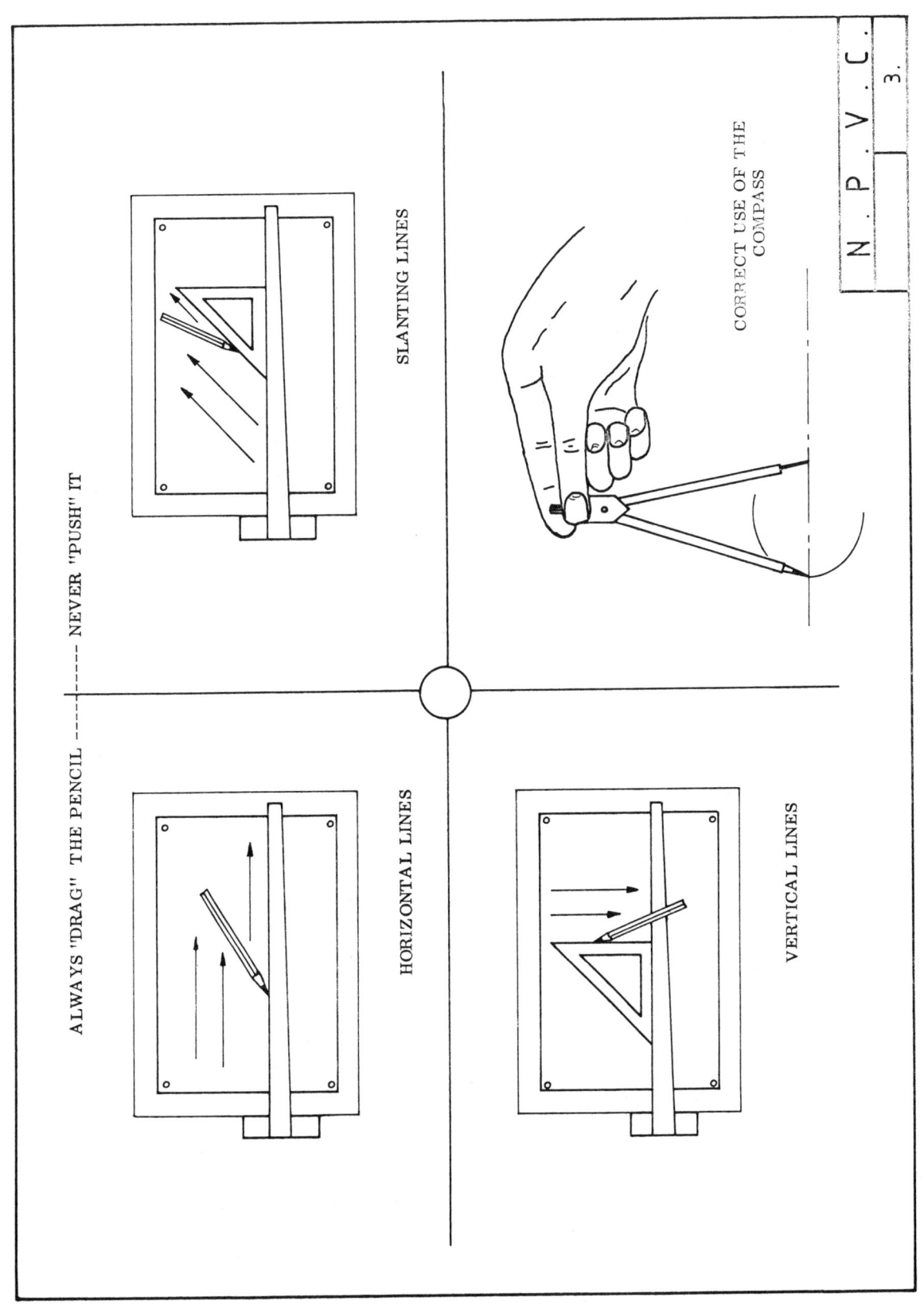

LETTERING

Any writing which is done on a drawing is always in the form of lettering and never in ordinary writing. Make sure that all the lettering on a drawing is the same height, with the exception of the title, which may be in larger capitals.

GUIDELINES

Guidelines are made to show the height and proper alignment of the letters on a drawing. They should be ruled very faintly with a sharp pencil so that they can be erased easily. For pencil lettering use an H or HB pencil.

TYPES OF LINES

Various types of lines are used on drawings; these are shown in the examples on the right.

A- FAINT LINES: These should stand out very fine and clear.
B- VISIBLE OUTLINES: These are bold continous lines and they should stand out clearly.
C- HIDDEN OUTLINES: These are chains of short, sharp lines.
D- SECTION LINES: These show the plane on which an object is cut for the section view.
E- DIMENSION LINES: These lines always have a number on them, giving the length of a part of the object.
F- CENTRE LINES: These show the centre line of an object and are usually used in sketching.
G- RADIUS AND DIAMETER LINES: In drawing circles, two lines are particularly important; these are the radius and diameter of the circle.

WHICH KIND OF LINES DO YOU SEE IN DRAWINGS 1, 2, AND 3?

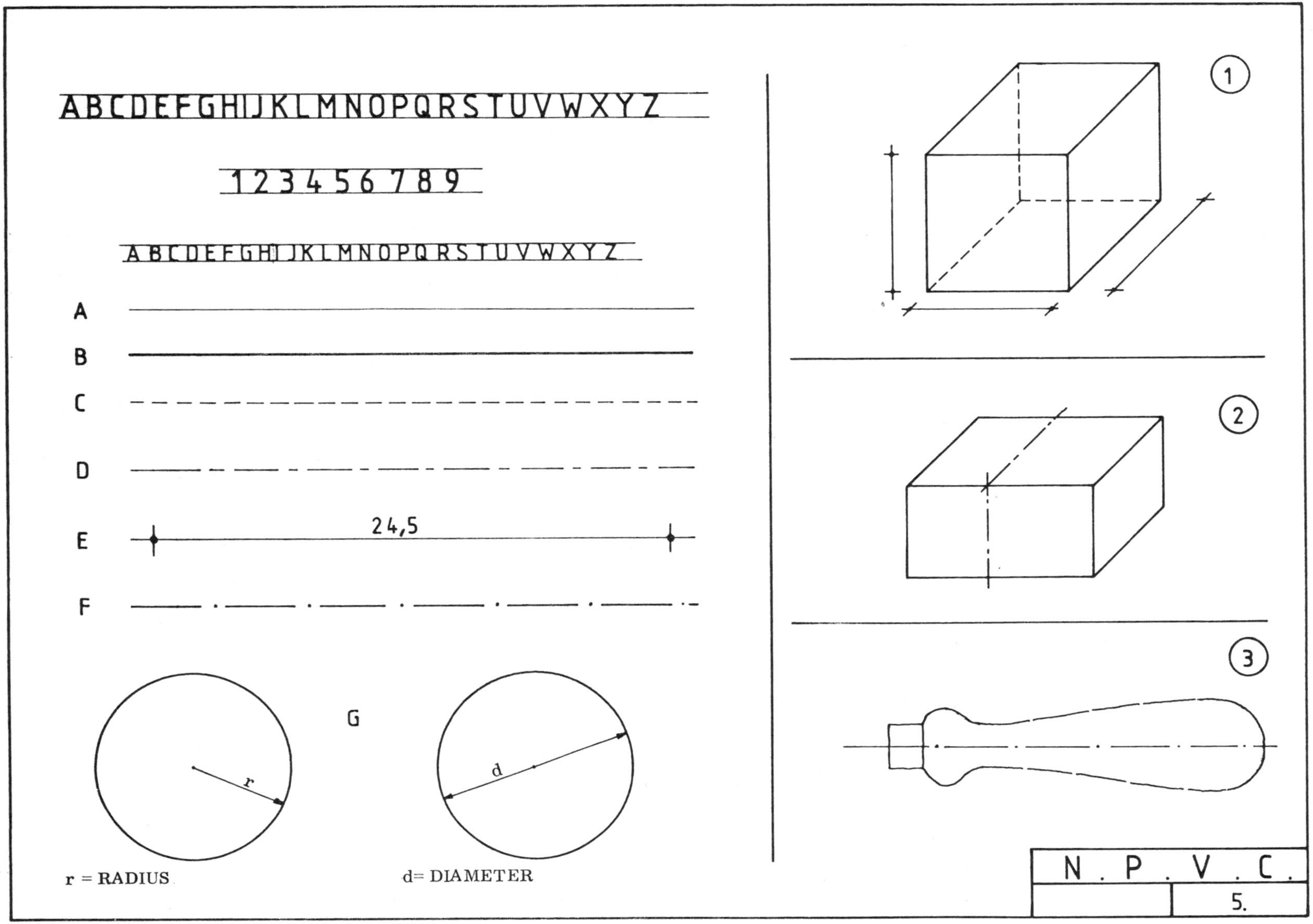

ABCDEFGHIJKLMNOPQRSTUVWXYZ

1234567890

ABCDEFGHIJKLMNOPQRSTUVWXYZ

A ─────────────
B ─────────────
C ─ ─ ─ ─ ─ ─ ─
D ─·─·─·─·─·─·─
E ●────24,5────●
F ─··─··─··─··─

r = RADIUS   G   d = DIAMETER

① ② ③

N. P. V. C.

5.

ORTHOGRAPHIC DRAWING

An orthographic drawing shows an object by means of a number of different views. Each view shows one side of the object as it is seen if looked at straight on. The diagram here shows a rectangular solid (A) along with an orthographic drawing of it (B).

- If a certain view will be the same as another view, you need draw only one of the views (compare the left side view and right side view from A).

- If certain measurements on one view are the same as measurements on another view, you should not label these measurements on both views. It is better to arrange the views so it is clear that the measurements are the same. The drawings on the right show how this is done (B & D).

In making an orthographic drawing, it is necessary to first choose the "front view". Any side can be choosen as the front view (even the top!) although it is usual to choose the most important side of the object.

- THE IMPORTANT POINTS TO REMEMBER IN MAKING AN ORTHOGRAPHIC DRAWING ARE:

- Space the views an even distance apart.
- Make accurate measurements.
- Make clear lines. Make sure that the outlines are darker than the dimension lines.
- The scale, in cm or mm, should be mentioned in the title block of the drawing (see page 9).
- The lettering must be uniform and clear.

- STUDY: Look at solid A and orthographic drawing B. You need only three views for the drawing. Why is this? Look at solid C and orthographic drawing D. Why do you need more views in orthographic drawing D?

- DRAW: Make an orthographic drawing of solid C with side X as the front view.

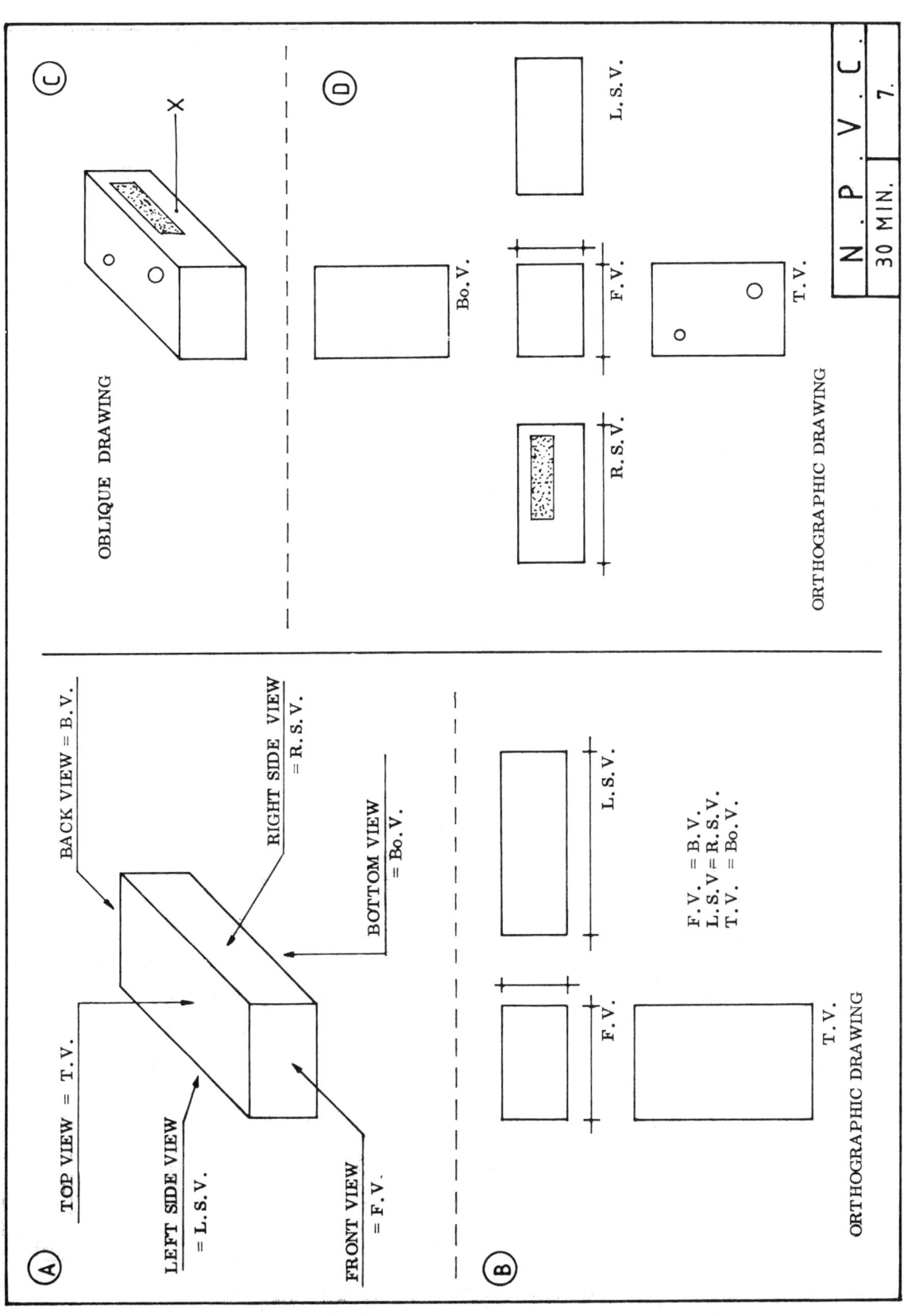

SETTING OUT THE DRAWING

When you set out the different views of an object on the paper, make sure that they are evenly spaced. Never squash them all on one side. Leave enough space for the necessary titles, sub-titles, and descriptions.

The name of each view should be written on the lower right side of the view. Abbreviations of the view names can be used:

F.V.	= front view	B.V.	= back view		
T.V.	= top view	Bo.V.	= bottom view		
L.S.V.	= left side view	R.S.V.	= right side view		

MARGIN LINE

A margin line should be drawn around the paper, 1 cm from the edge.

DIMENSIONS

No drawing is complete unless the dimensions or lengths of all the sides are given. The illustration on the right shows how dimensions are given on a drawing. Make sure that no dimensions are left out or repeated.

TITLE BLOCK

When the drawing is complete, the last thing that has to be done is to make a title block in the bottom right corner of the paper. The title block gives the title of the drawing, the scale used, the date on which it was drawn, and who drew it. The standard size of the title block is given in the drawing on the right.

- DRAWING: Make an orthographic drawing of solid A.

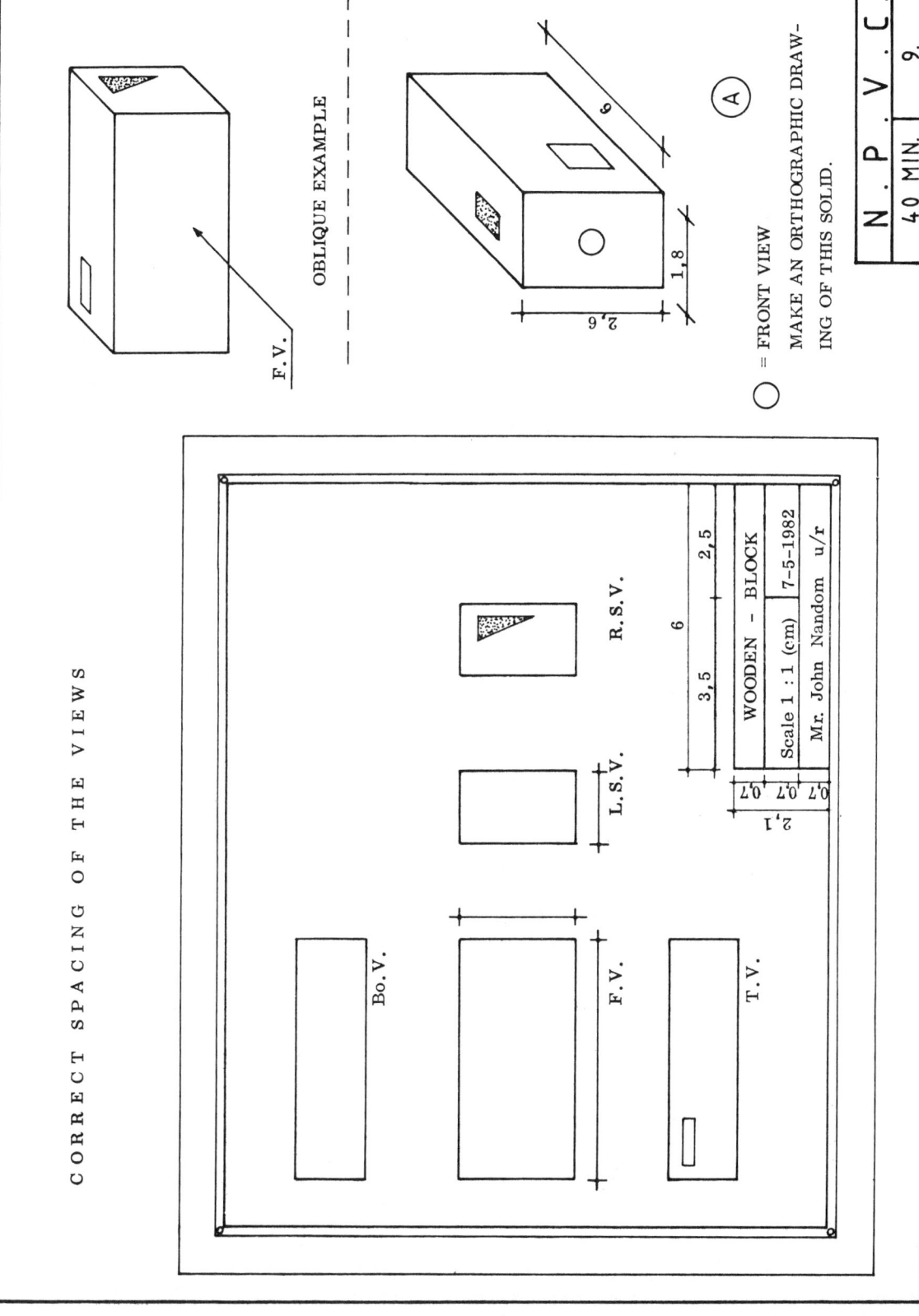

OBLIQUE DRAWING

An oblique drawing is a pictorial representation of an object. The difference between oblique and orthographic drawings is that in oblique drawings the object appears as a single drawing and looks like it does in real life. It is important that you be able to understand both types of drawing and be able to turn one type of drawing into the other.

In oblique drawing one face of the object is drawn in its true shape. This face is always the front view. The rest of the drawing is then built up using three "axes" or directions. One axis is always horizontal (A); another axis is always vertical (B); but the third axis (C) can be at any other angle. It is usual, however, to have the third axis at 45 degrees. These are shown in the illustration on the right.

Oblique drawing can be done step by step.

1 - Choose the front view. Draw it faintly on your paper.
2 - Draw the oblique lines at 45 degrees. Make sure that they are all the same length.
3 - Connect the oblique lines together using vertical lines or horizontal lines as needed.
4 - Erase all unnecessary lines.
5 - Mark all dimensions.
6 - Redraw the outline of the object with an HB pencil to make it darker.

CURVED OBJECTS

An object with a curved surface can easily be shown in an oblique drawing. To do this, draw the shape of the curve in the front view, so that the curve appears in its true shape. Finish off by drawing the oblique lines etc. as usual (see below).

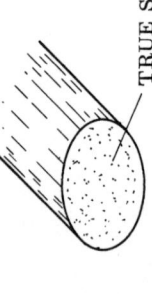
TRUE SHAPE

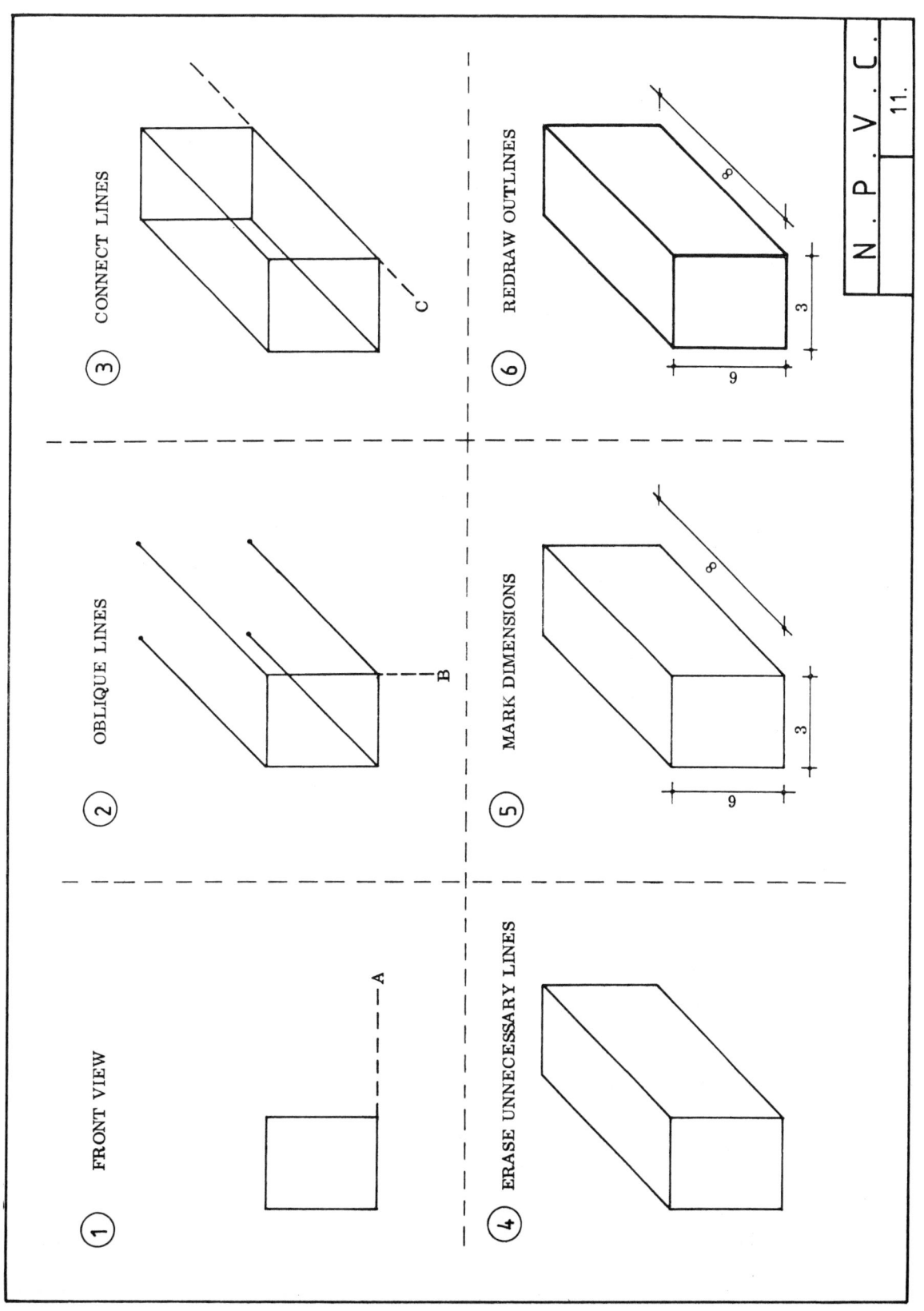

TECHNIQUES IN OBLIQUE DRAWING

There are some techniques and conventions that you need to know to make good oblique drawings. The most important of these is choosing the best view of the object. The best view is the one that gives you the most information about the object. A good rule to follow is: the best view is the one that is the hardest to draw.

On the next page you can see four different drawings of the same object. They all have the same front view but they are all projected in different directions. You should see that they all give different amounts of information about the object (A).

- Which view gives you the most information?
- Which view is the hardest to draw?

The answer is "B", for both questions.

- CONVENTIONS: All types of drawing have their own conventions. At first the conventions may seem unnecessary, but you will soon find that they all help you to understand a drawing better. One of the most important conventions in oblique drawing is the marking of dimensions. The drawing here (B) shows how this should be done. Remember that the dimensions are given simply as numbers on the drawing itself, and that the unit (cm, mm, etc.) is given in the title block.

For example: if the length of a block is 6 cm, the dimension in the drawing is given as "6"; and the unit "cm" is recorded in the title block.

- DRAW: Draw or sketch several solids, as in the following pages. Judge which side should be the front view. Decide upon the position in which the solid will be drawn.

A FOUR DIFFERENT POSITIONS IN OBLIQUE

T.V./F.V./L.S.V.

F.V./L.S.V./Bo.V.

T.V./F.V./R.S.V.

F.V./R.S.V./Bo.V.

WHICH POSITION IS THE BEST?

MARK DIMENSIONS LIKE THIS !

B

F.V./Bo.V./R.S.V.

SECTIONS
TOTAL

N.P.V.C.

13.

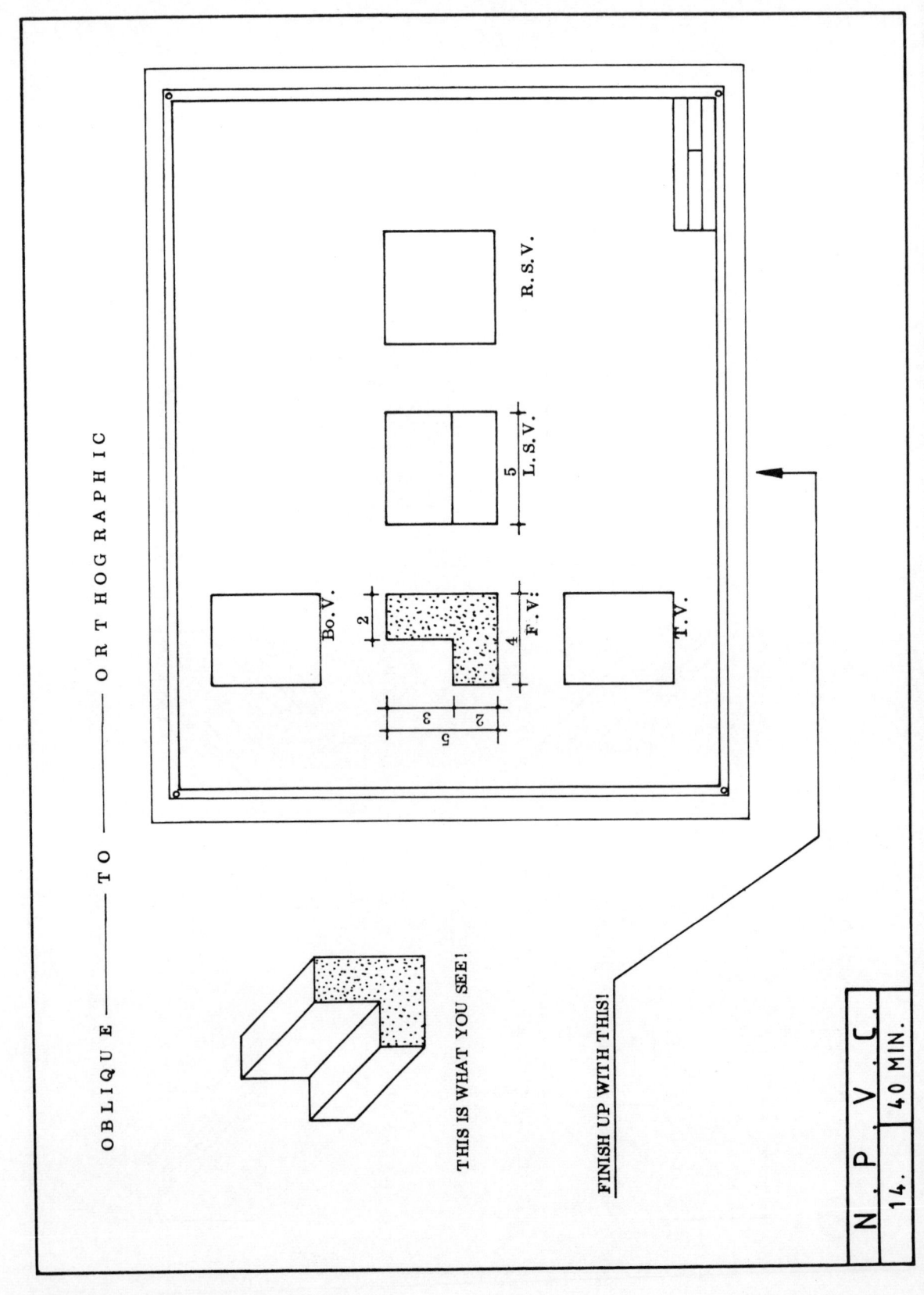

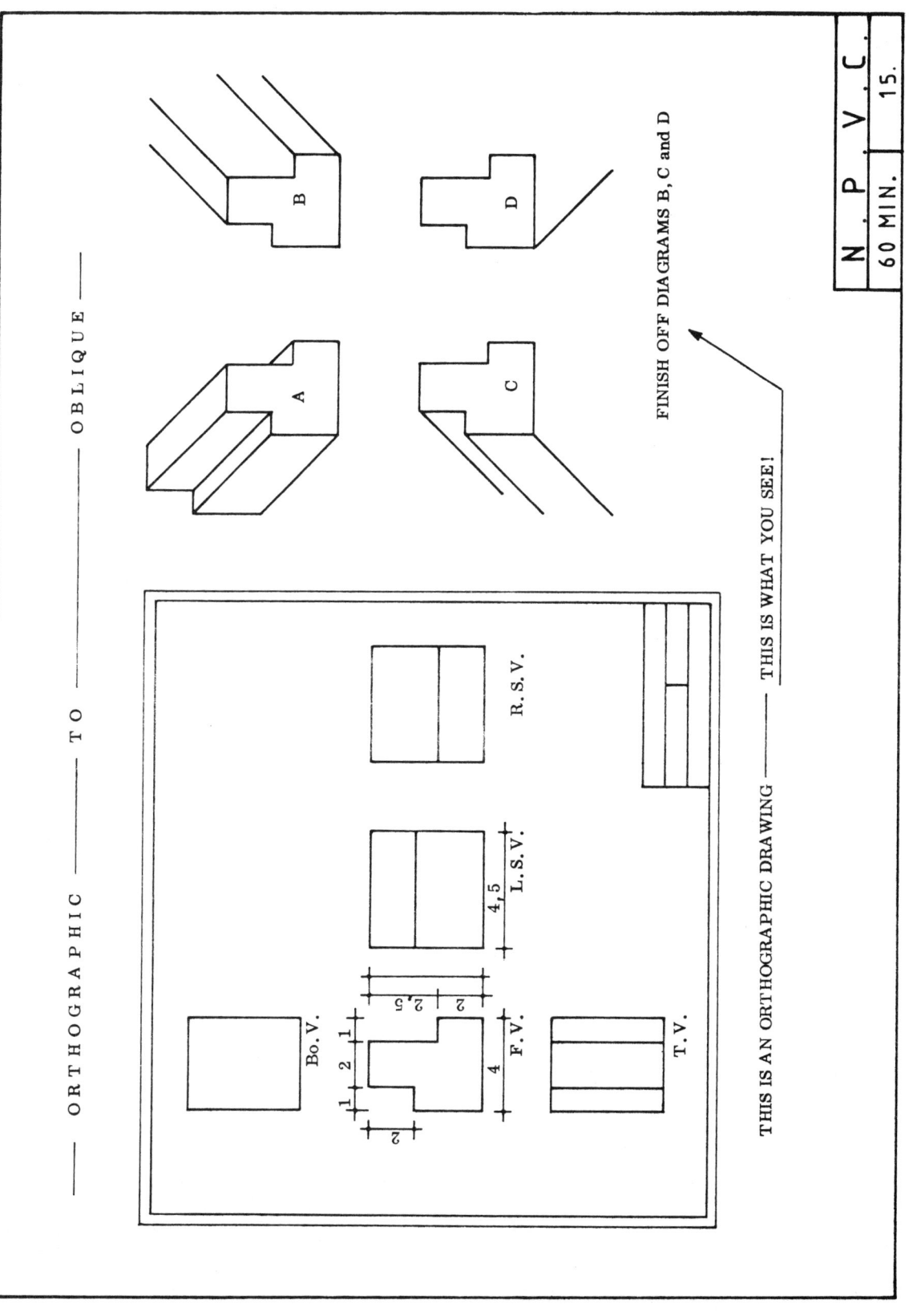

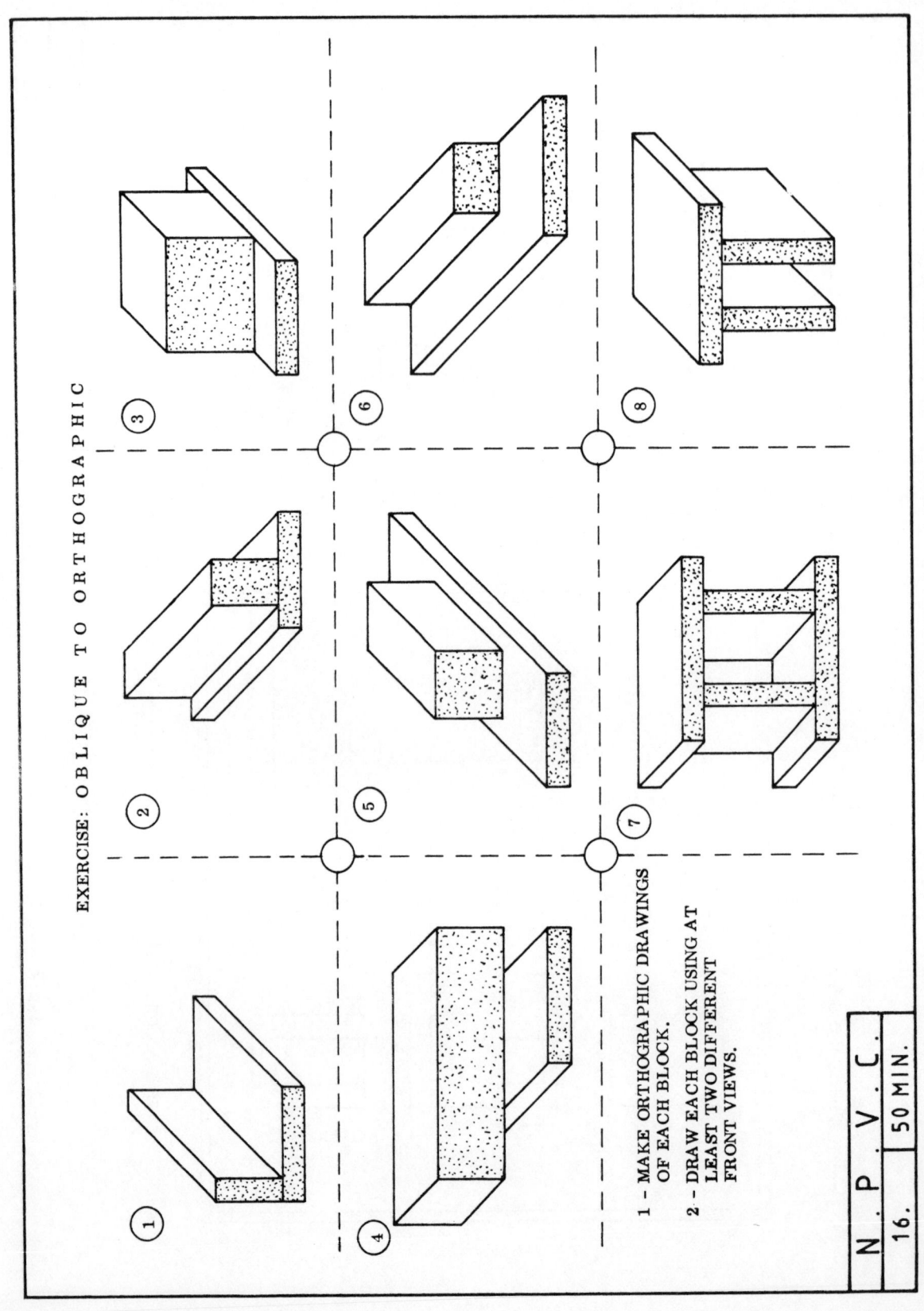

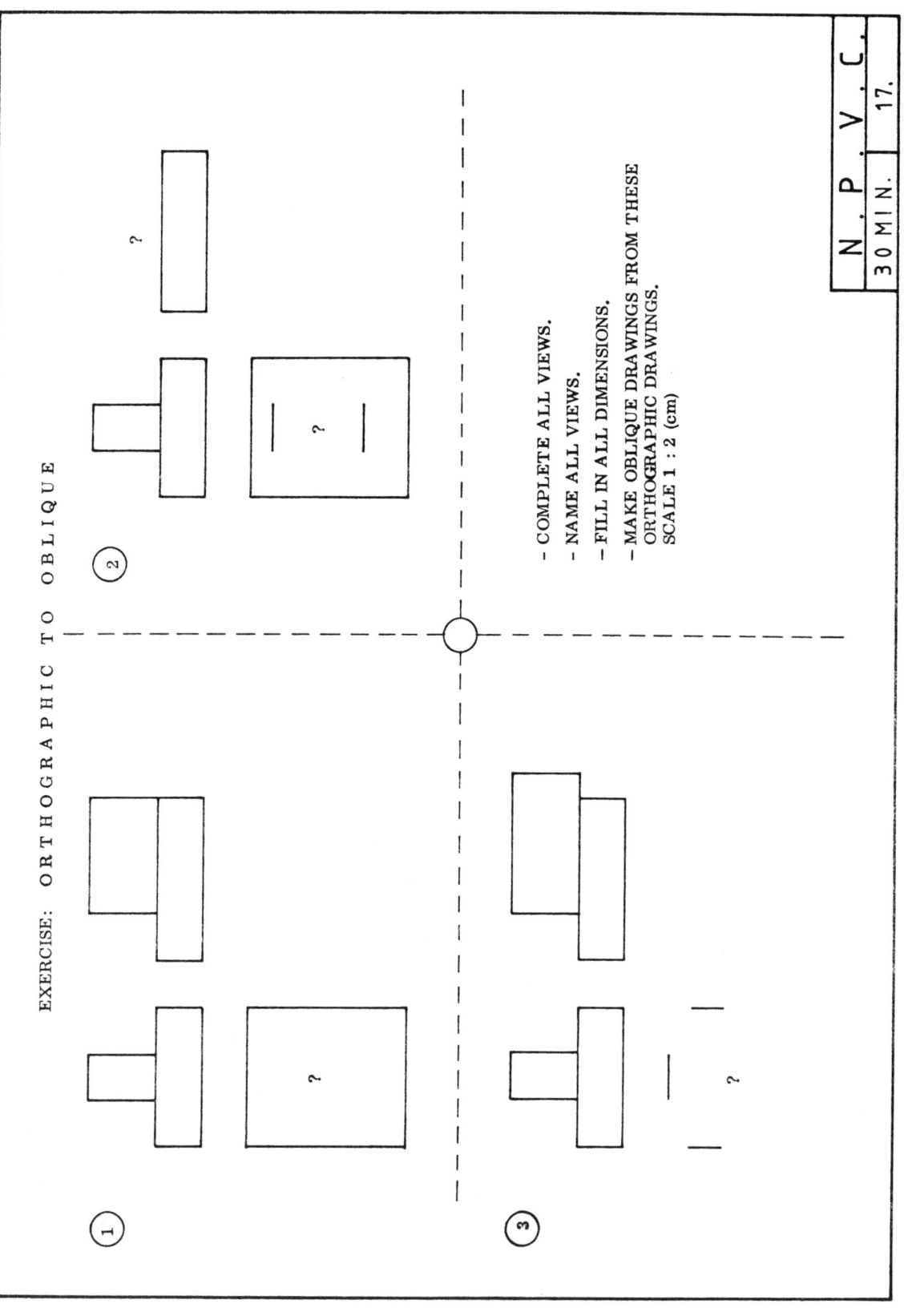

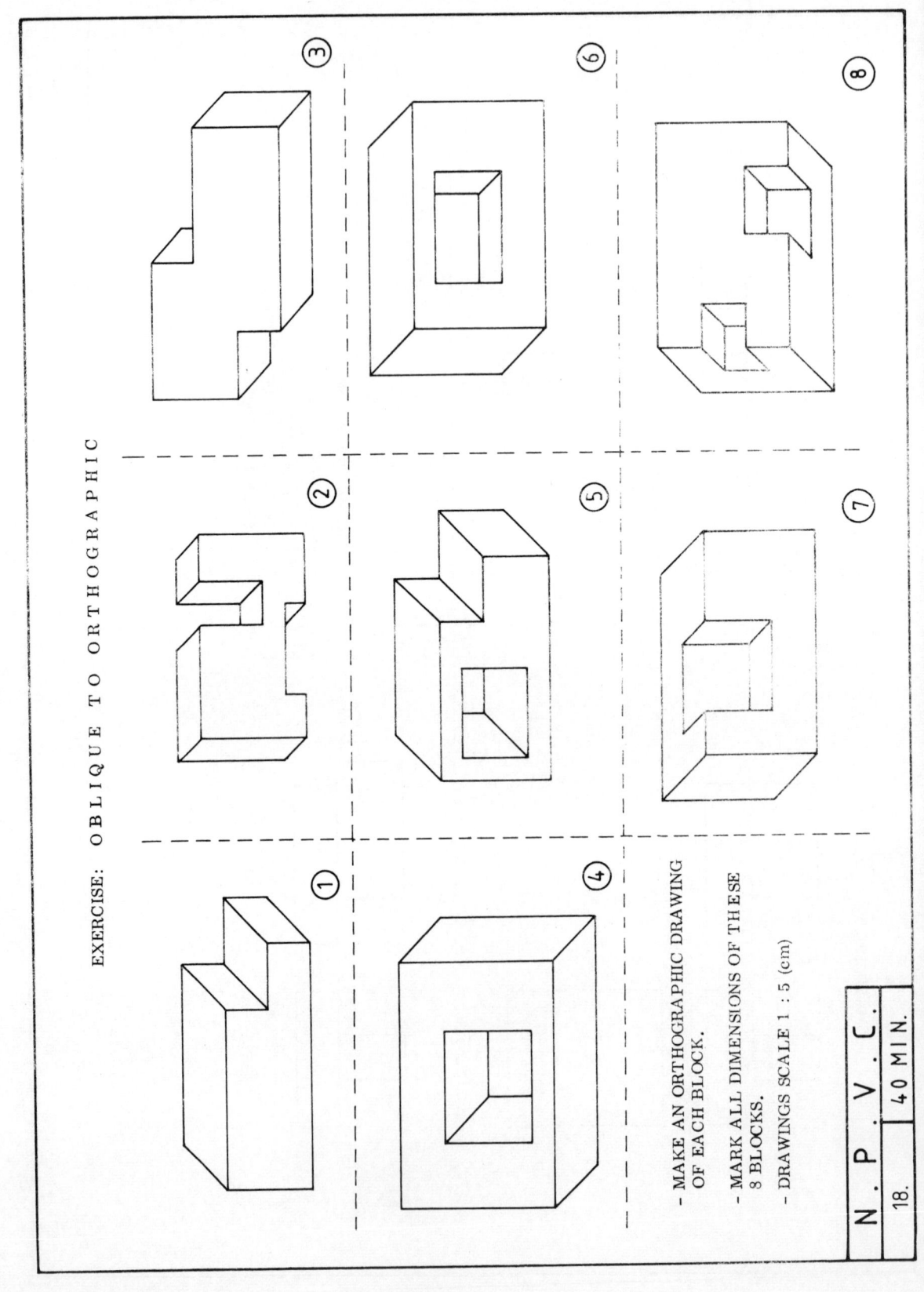

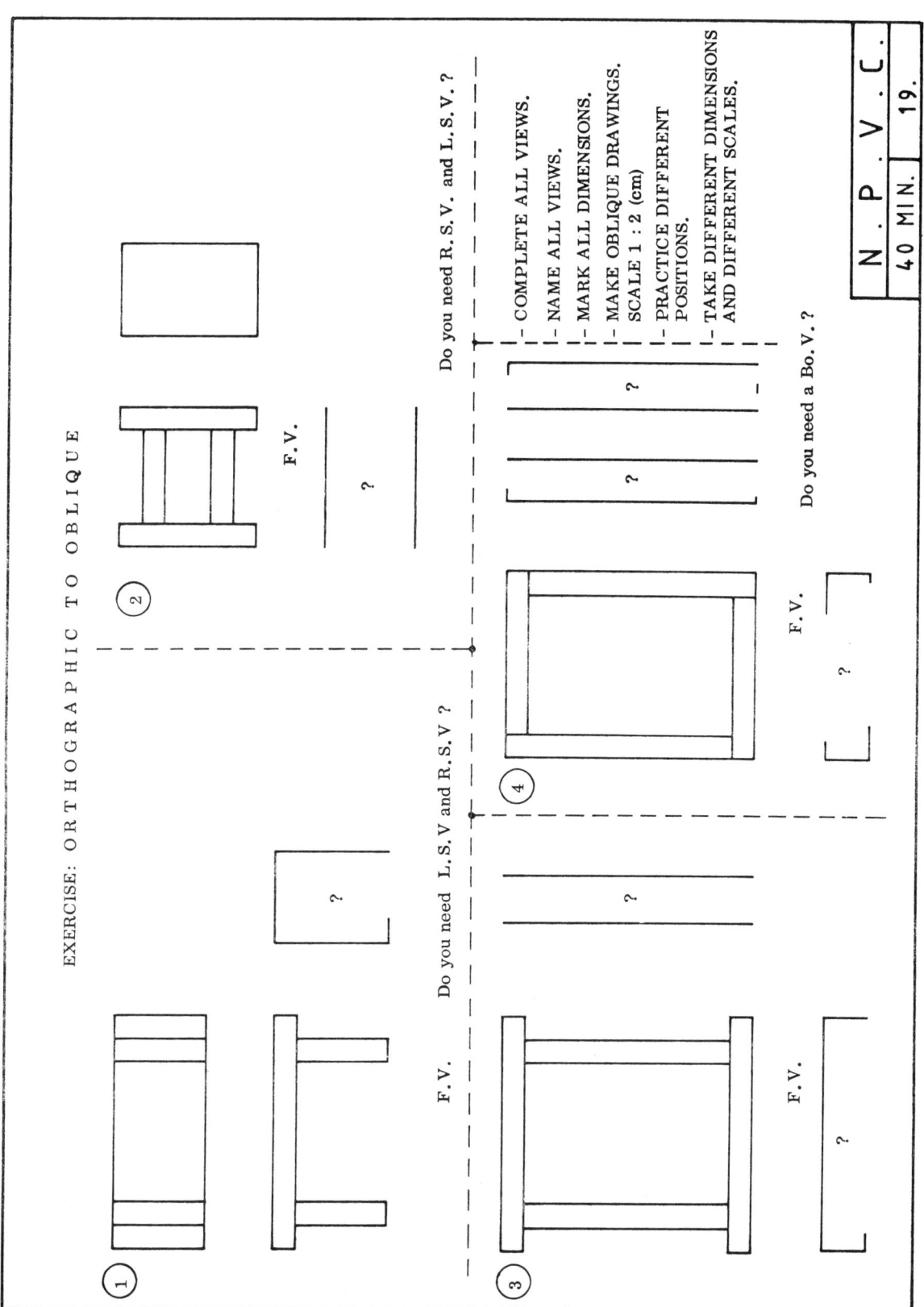

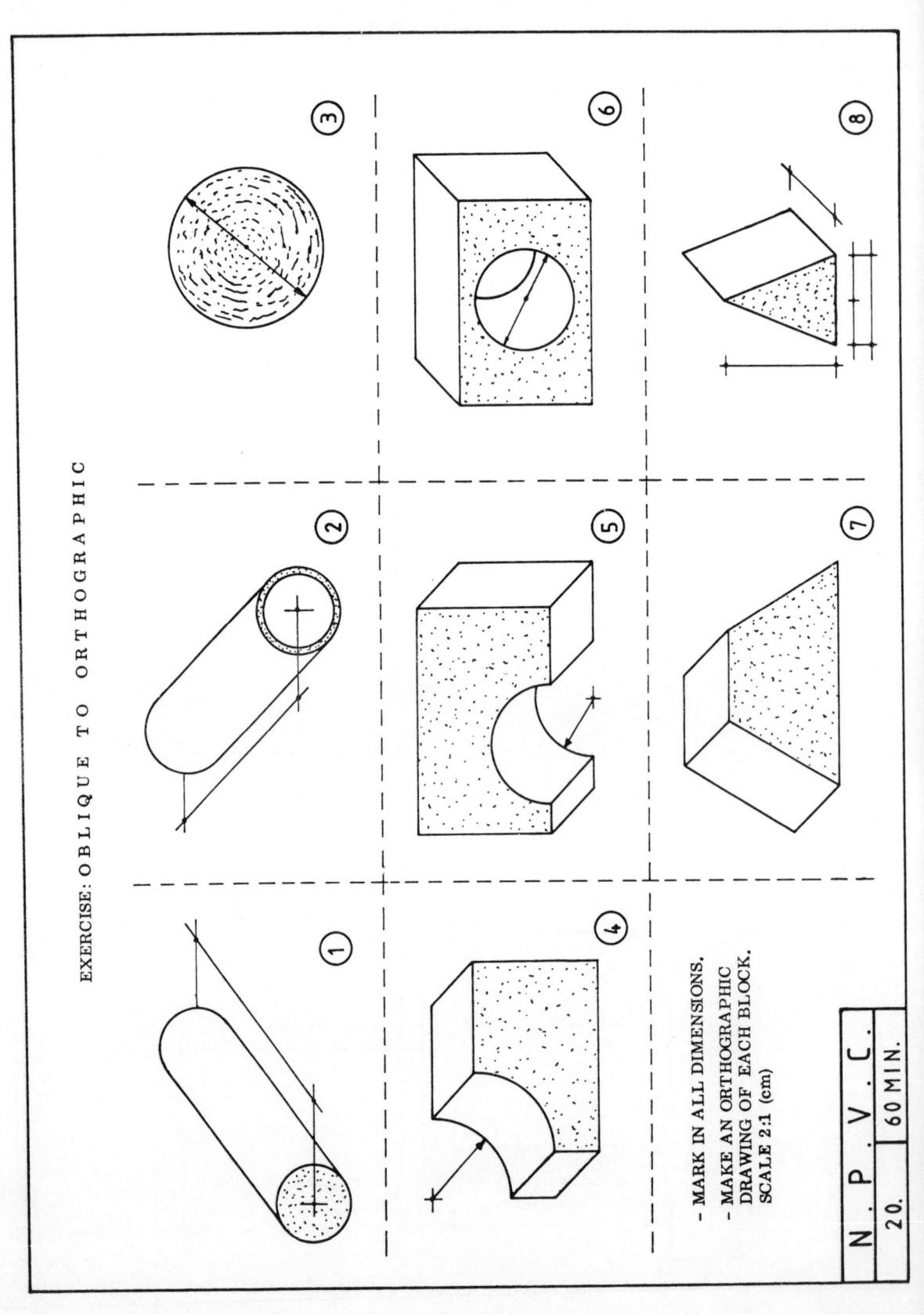

EXERCISE: ORTHOGRAPHIC TO OBLIQUE

- NAME ALL VIEWS.
- MARK ALL DIMENSIONS.
- MAKE OBLIQUE DRAWINGS

1: FV-TV-LSV   2: FV-BoV-RSV
3: FV-TV-RSV   4: FV-TV-RSV
5: FV-BoV-LSV   6: FV-BoV-RSV
7: FV-BoV-LSV   8: FV-TV-RSV

N.	P.	V.	C.
40 MIN.			21.

SCALE DRAWING

Before you start to make any kind of workpiece, it is necessary to make a drawing showing how it should be made. This drawing is usually called the "layout". When making a layout it is important to use the correct scale. The scale tells you how much bigger or smaller the actual object is, compared to the drawing.

- Large objects have to be drawn smaller than they actually are: in this case you have to use a REDUCED SCALE.
- Very small objects have to be drawn larger than they actually are so that all the details can be seen. In this case you have to use an ENLARGED SCALE.
- If the object is neither very large nor very small, it may be drawn as it actually is. In this case you use FULL SCALE.

SCALES

The scale affects the size of every side of the object in the drawing. Make sure that you draw all lengths, widths, and to the correct scale.

- EXAMPLES OF SCALE: A scale of 1 : 5 (mm) tells you that 1 mm on the drawing represents 5 mm in real life. In other words, the drawing shows the object as 1/5th of its real size, with all dimensions in millimetres.

A scale of 2 : 1 (cm) tells you that 2 cm on the drawing represents 1 cm in real life. In other words, the drawing shows the object as twice its actual size, and all the dimensions are in centimetres.

What do the scales 1 : 10 (m) and 5 : 1 (cm) tell you?

- COMMON SCALES: Some commonly used scales are 1 : 2; 1 : 5; 1 : 10; 1 : 20; and 1 : 50.

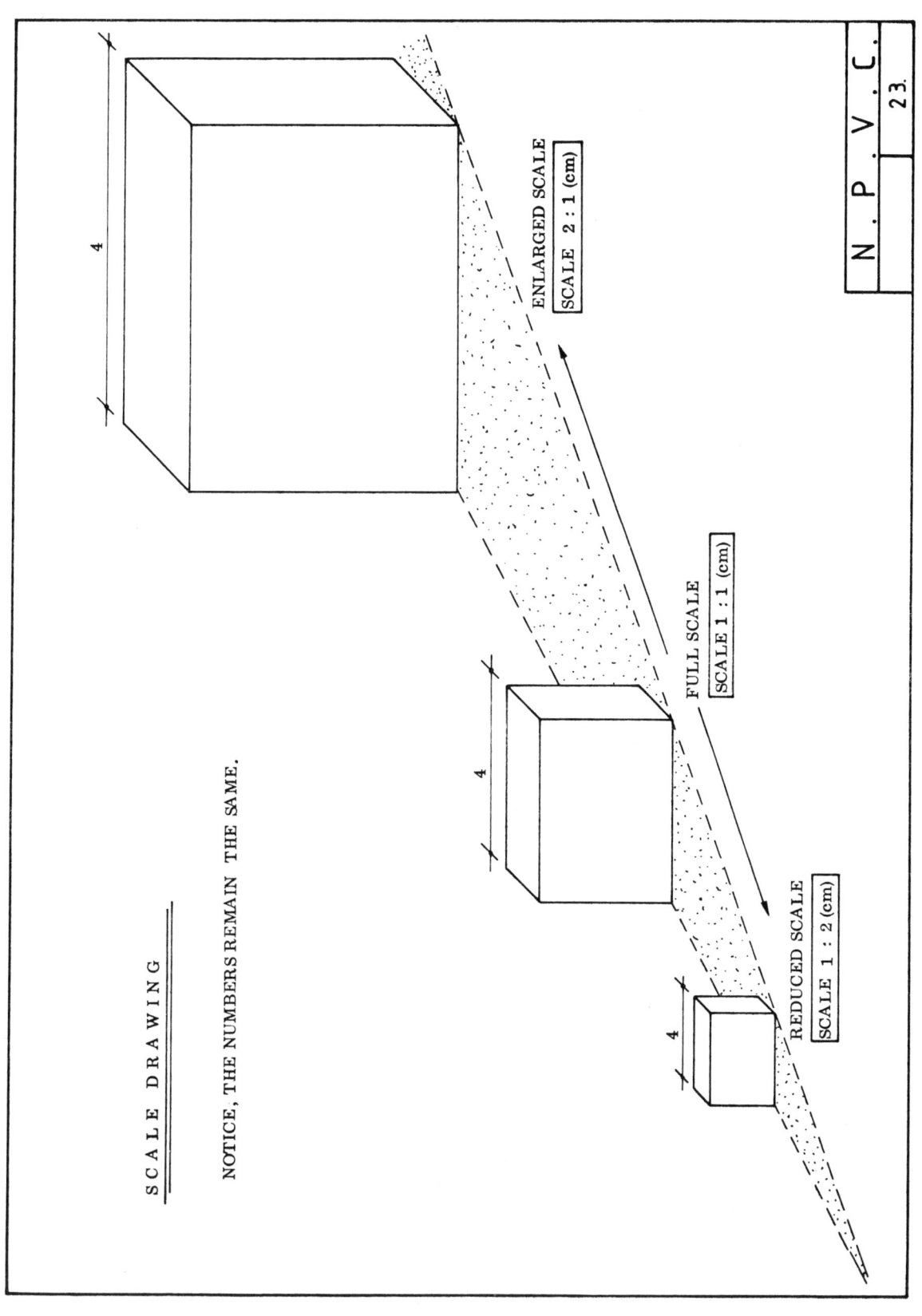

INSIDE AND OUTSIDE DIMENSIONS

If you measure the inside of a box, you will find the measurements are smaller than the outside measurements. This is because of the thicknesses of the sides of the box. Look at the drawing on the next page (A) and note that the outside length of the box is equal to the inside length plus TWICE the thickness of the walls. The same applies to houses: the outside dimensions of a room will always be larger than the inside dimensions.

— EXAMPLE: A box has inside dimensions of 30 x 55 cm and is made of boards which are 3 cm thick. What are the outside dimensions of this box? (see A).

FRAMES

Many articles such as windows and doors can be bought as ready made pieces. These have certain sizes. Both doors and windows usually need a frame around them, and this has to be taken into account when the plans are made for the house.

Figs. B and C on the next page show some ready made articles. Below each you see the same article with a frame around it. In both cases the frame is 6 cm thick.

Fig. B shows a single article such as a casement. The frame goes all the way around the casement, which means that the window opening in the wall must be 12 cm longer and 12 cm higher than the casement itself, so that there is room for the frame.

Fig. C shows a pair of casements. In this case the frame goes all the way around the casements and also between them. This means that the window opening in the wall must be 18 cm longer and 12 cm higher than the casement itself.

EXERCISE: IN AND OUTSIDE DIMENSIONS

A - THE BOX HAS WALLS THAT ARE 3 cm THICK.
MARK IN THE OUTSIDE DIMENSIONS.

B - MARK IN ALL DIMENSIONS.

C - MARK IN ALL DIMENSIONS.

N.P.V.C.
25.

DESIGNING FROM SKETCHES

You will often have people coming to you with a rough sketch of something they want you to make, for example a set of casements. In order to make the frame around the casements and to know the size of the opening which has to be made in the wall, there are two important things to notice about most rough sketches:

- The dimensions are the <u>outside</u> dimensions of the casements.
- The thickness of the frame has been left out.

Someone may come to you with a rough sketch of a complicated box with many partitions. In order to be able to construct the box, you have to notice two important things about the sketch:

- The dimensions are the <u>inside</u> dimensions of the partitions.
- The thickness of the partitions and the sides of the box has been left out.

As a craftsman you have to turn these types of sketches into good plans. The plans you draw should have inside and outside dimensions, as well as the thicknesses of the members of the frame or partitions.

- BUILD UP YOUR PLAN IN THE FOLLOWING WAY:

  - Draw two planning lines, one vertical and one horizontal (lines A and B).
  - On line A, start at the left and mark the thickness of the frame member; then the width of the casement. Next mark the thickness of the central frame member and then the width of the right casement. Finally mark the thickness of the right frame member.
  - Starting from the bottom, do the same thing on line B. Your marks should indicate in order: frame; casement; frame; casement; frame; casement; frame.
  - Transfer lines A and B onto your drawing paper so that they form a right angle (page 28).
  - You should now be able to finish the drawing as shown on page 29.

In the original sketch here, it looks as though the large space at the top of the sketch was 259 cm long. On your final drawing you will see that it is 269 cm long. Can you see why?

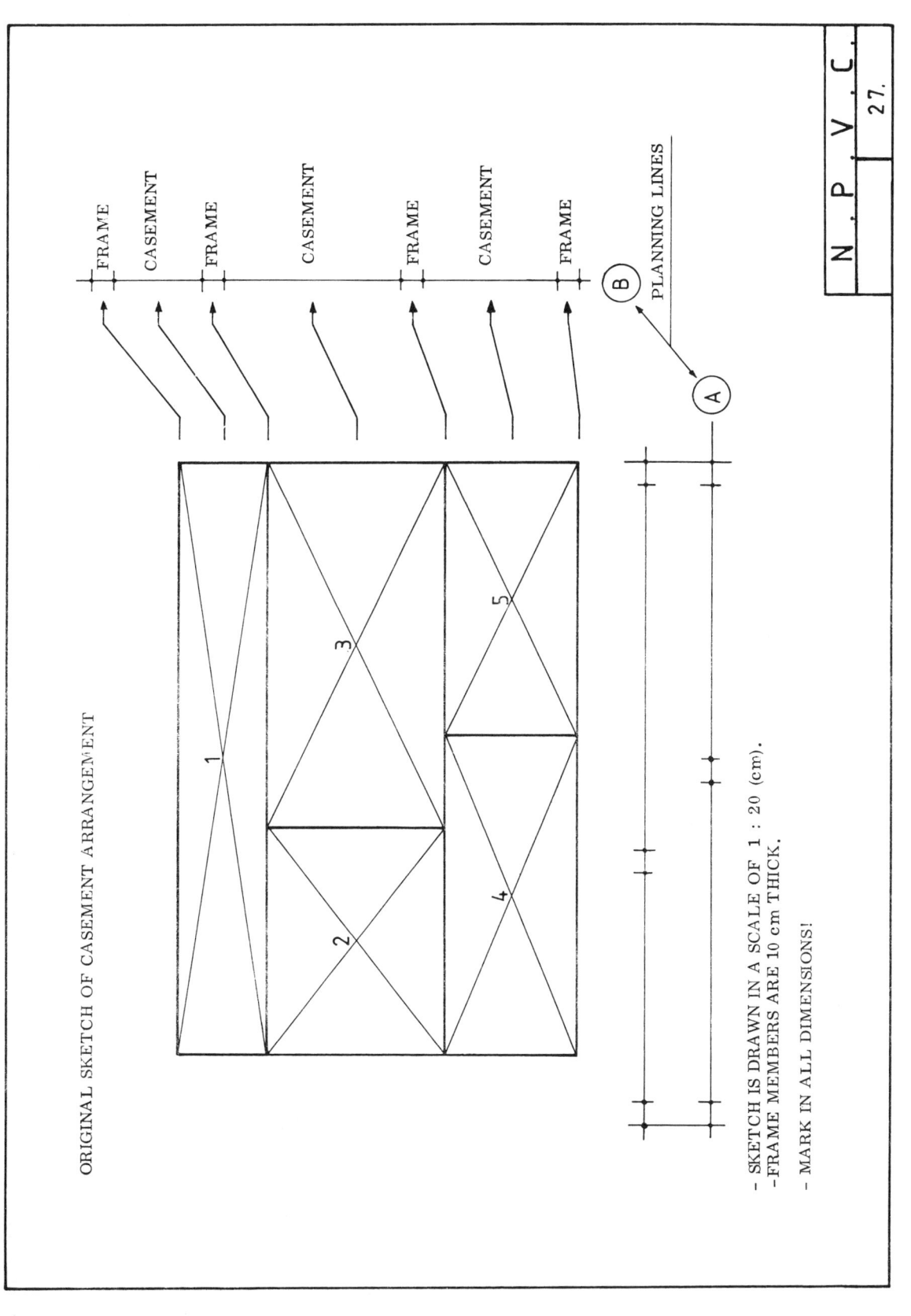

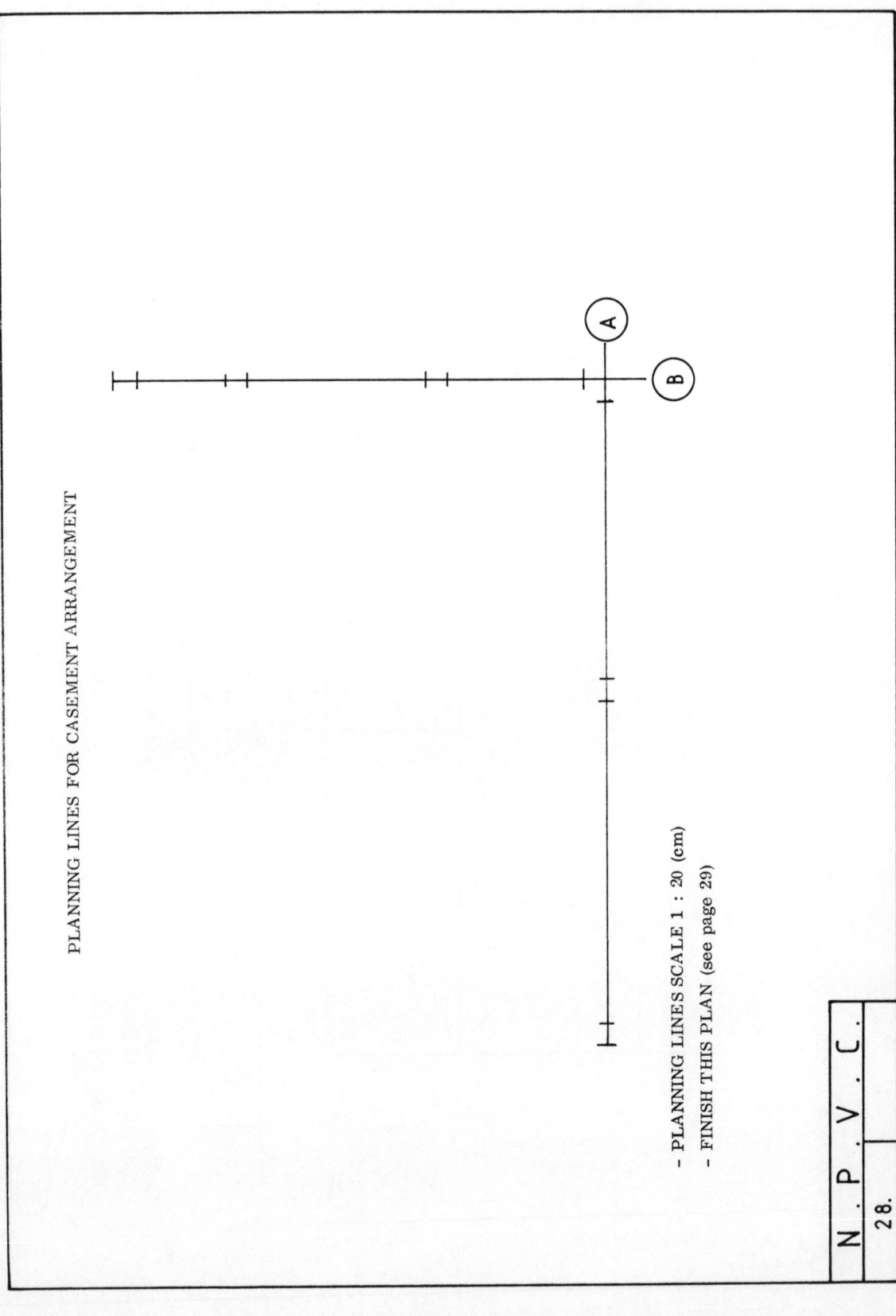

FINISHED PLAN OF CASEMENT ARRANGEMENT

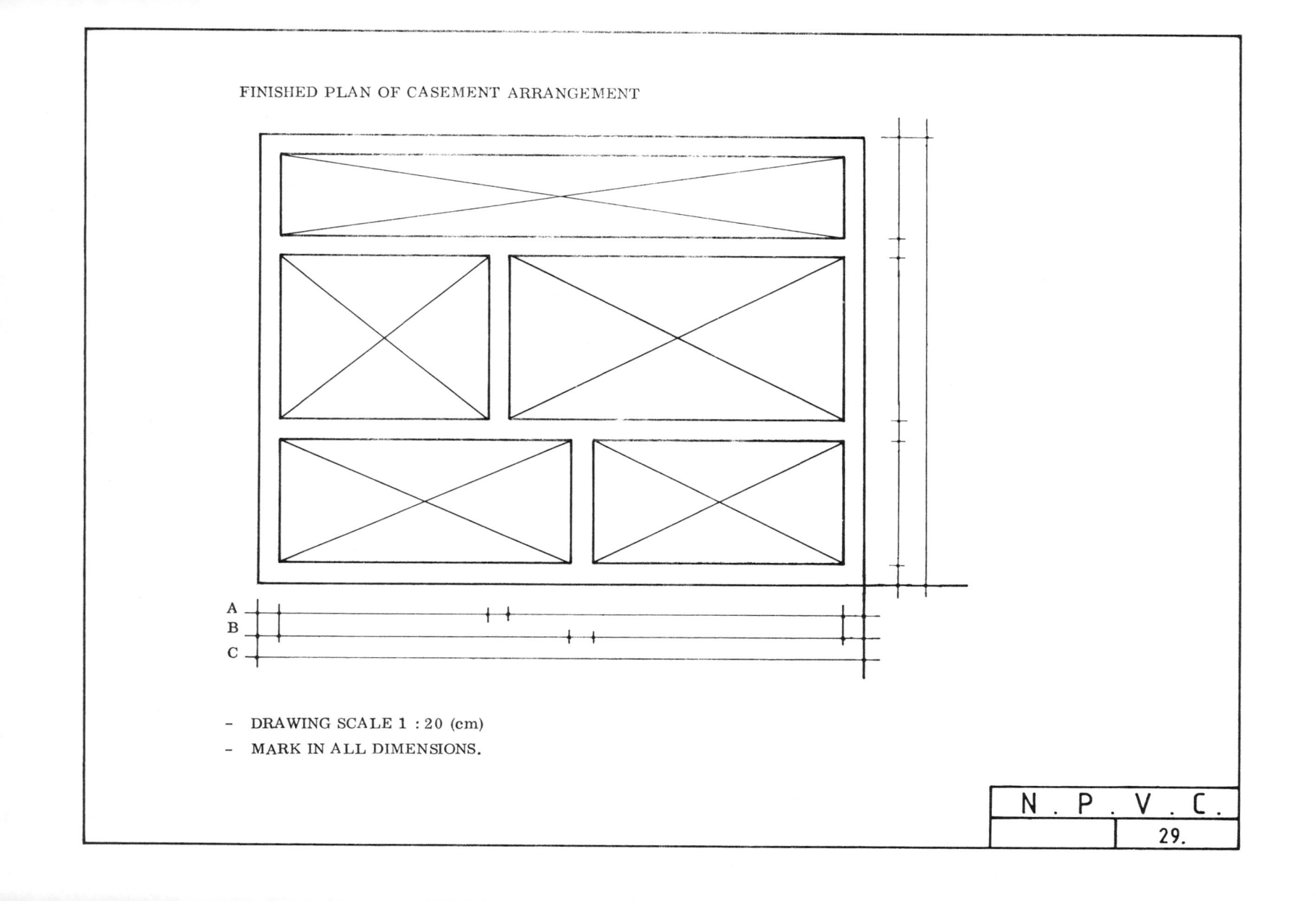

A
B
C

- DRAWING SCALE 1 : 20 (cm)
- MARK IN ALL DIMENSIONS.

N.P.V.C.

DESIGN

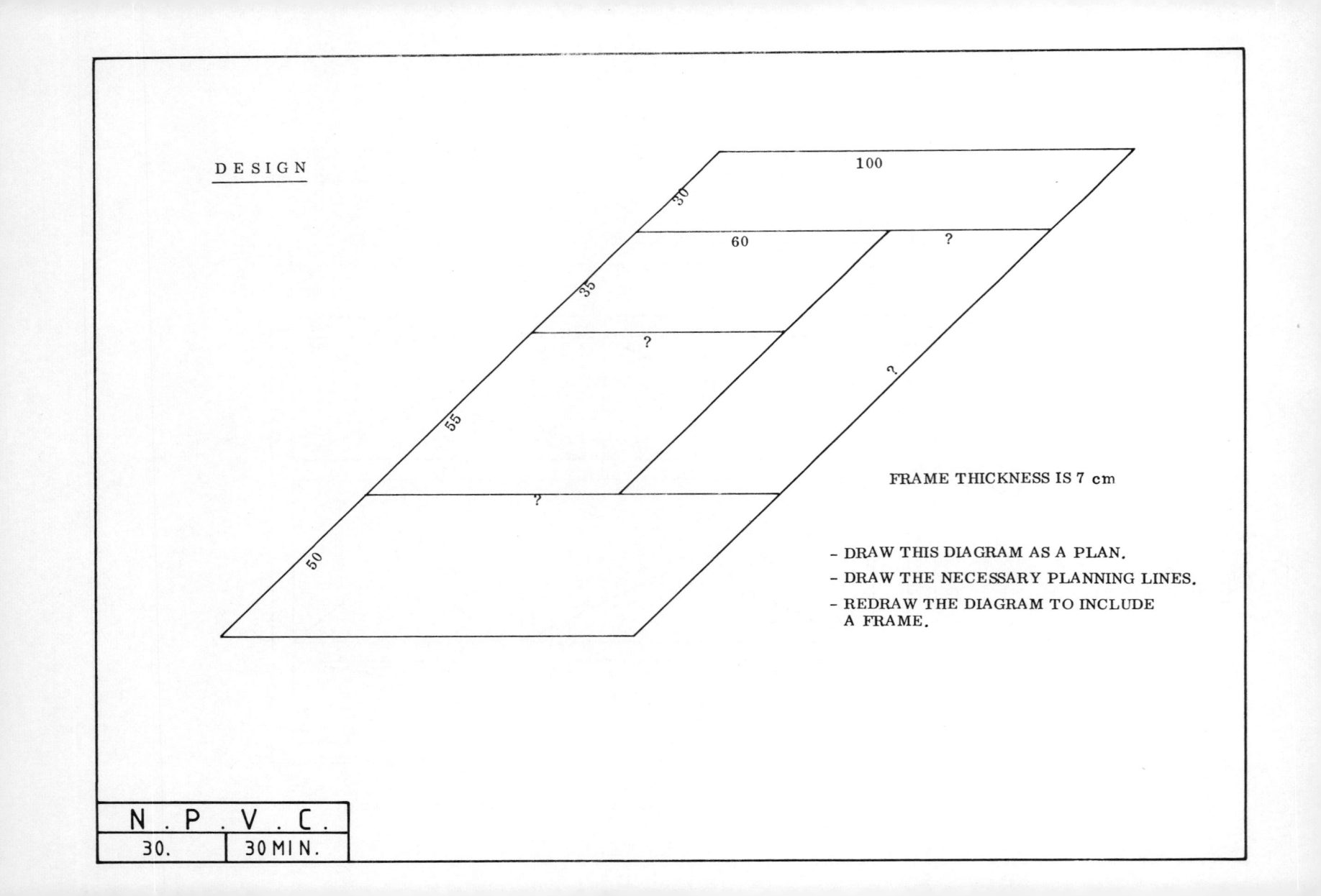

100
60
30
35
55
50
?
?
?

FRAME THICKNESS IS 7 cm

- DRAW THIS DIAGRAM AS A PLAN.
- DRAW THE NECESSARY PLANNING LINES.
- REDRAW THE DIAGRAM TO INCLUDE A FRAME.

N. P. V. C.	
30.	30 MIN.

DESIGN

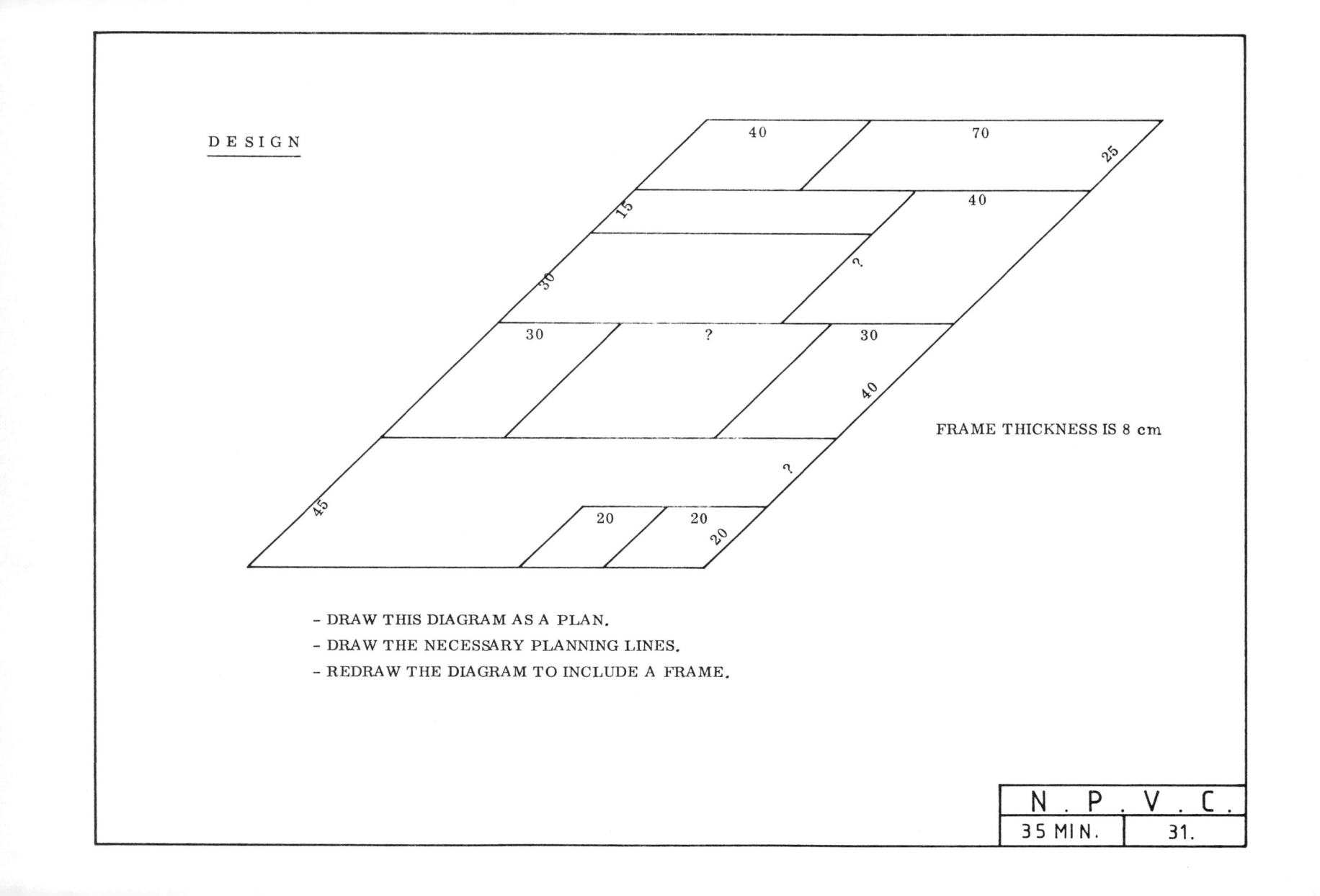

FRAME THICKNESS IS 8 cm

- DRAW THIS DIAGRAM AS A PLAN.
- DRAW THE NECESSARY PLANNING LINES.
- REDRAW THE DIAGRAM TO INCLUDE A FRAME.

N. P. V. C.	31.
35 MIN.	

① OBLIQUE DRAWING OF A BOX-LIKE OBJECT

B.L.
TOTAL HEIGHT
HEIGHT DIMENSION LINES
BOTTOM LEVEL = B.L.

A B CD

②

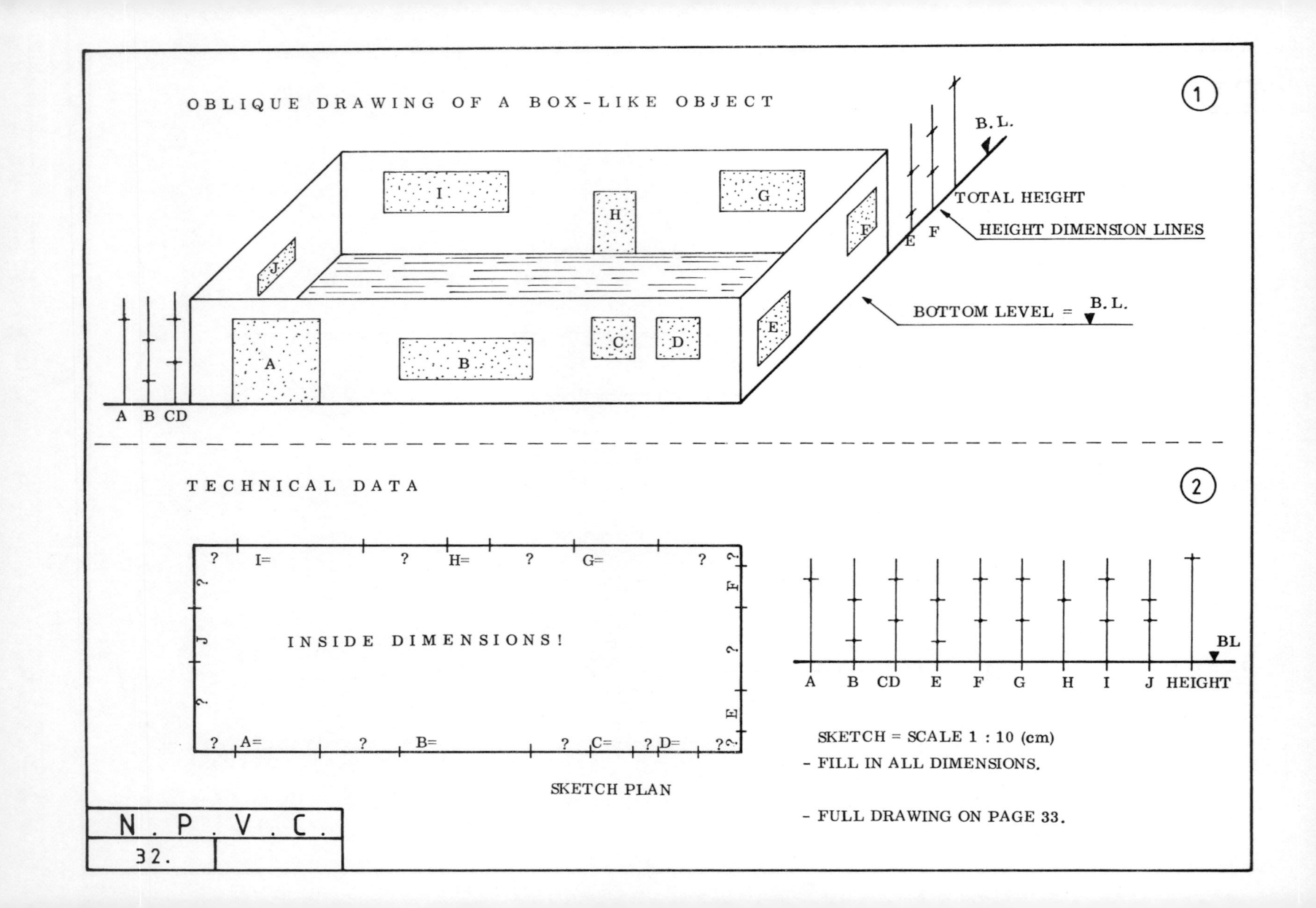

SKETCH = SCALE 1 : 10 (cm)
- FILL IN ALL DIMENSIONS.
- FULL DRAWING ON PAGE 33.

A B CD E F G H I J HEIGHT BL

TECHNICAL DATA

INSIDE DIMENSIONS!

I= ?   H= ?   G= ?   ?   ?   F ?
?   ?
?   ?
?   ?
A= ?   B= ?   ?   C= ? D= ?   E ?

SKETCH PLAN

N.P.V.C.
32.

ORTHOGRAPHIC DRAWING

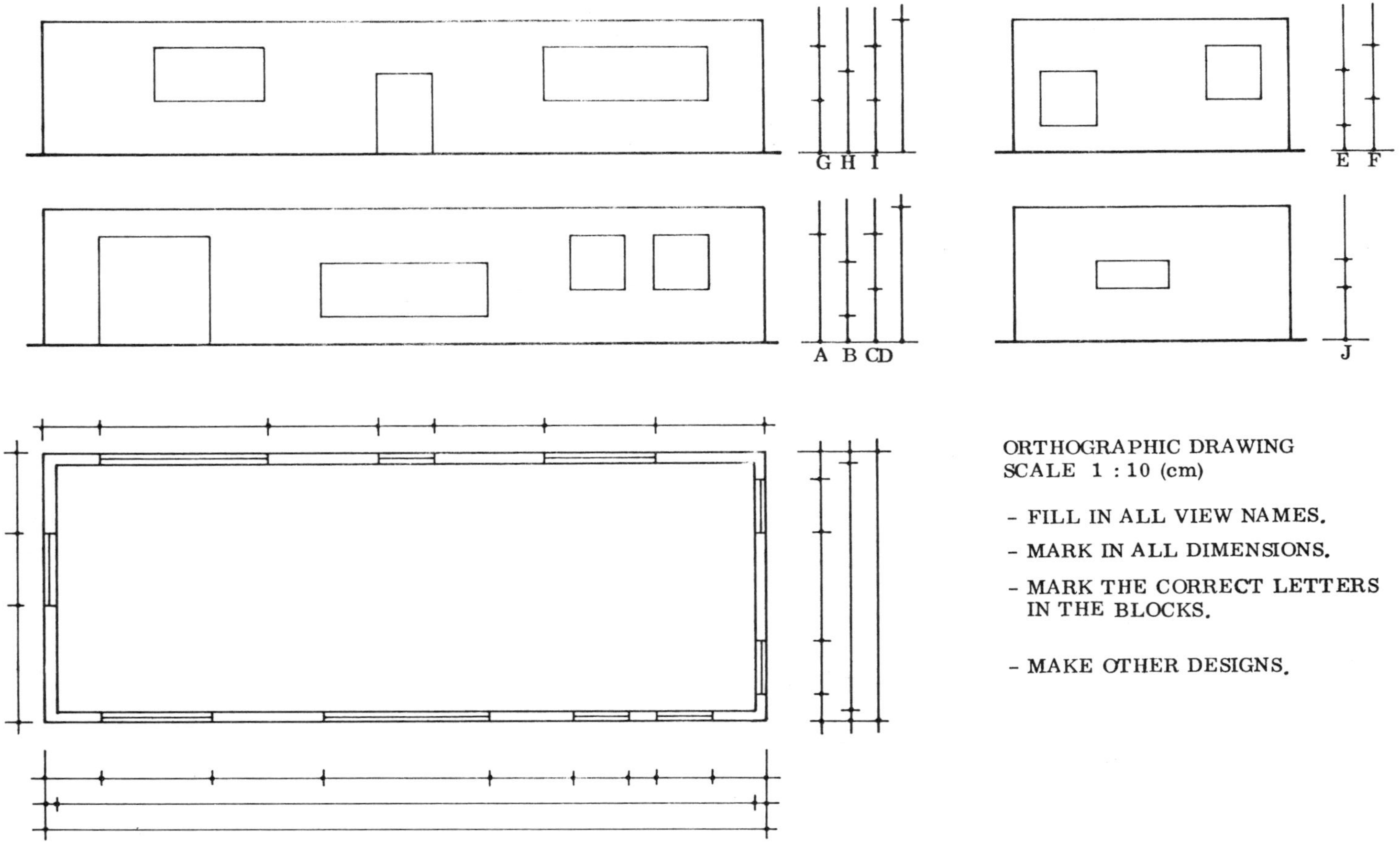

ORTHOGRAPHIC DRAWING
SCALE 1 : 10 (cm)

- FILL IN ALL VIEW NAMES.
- MARK IN ALL DIMENSIONS.
- MARK THE CORRECT LETTERS IN THE BLOCKS.
- MAKE OTHER DESIGNS.

N. P. V. C.

2 HOURS | 33.

CROSS SECTIONS

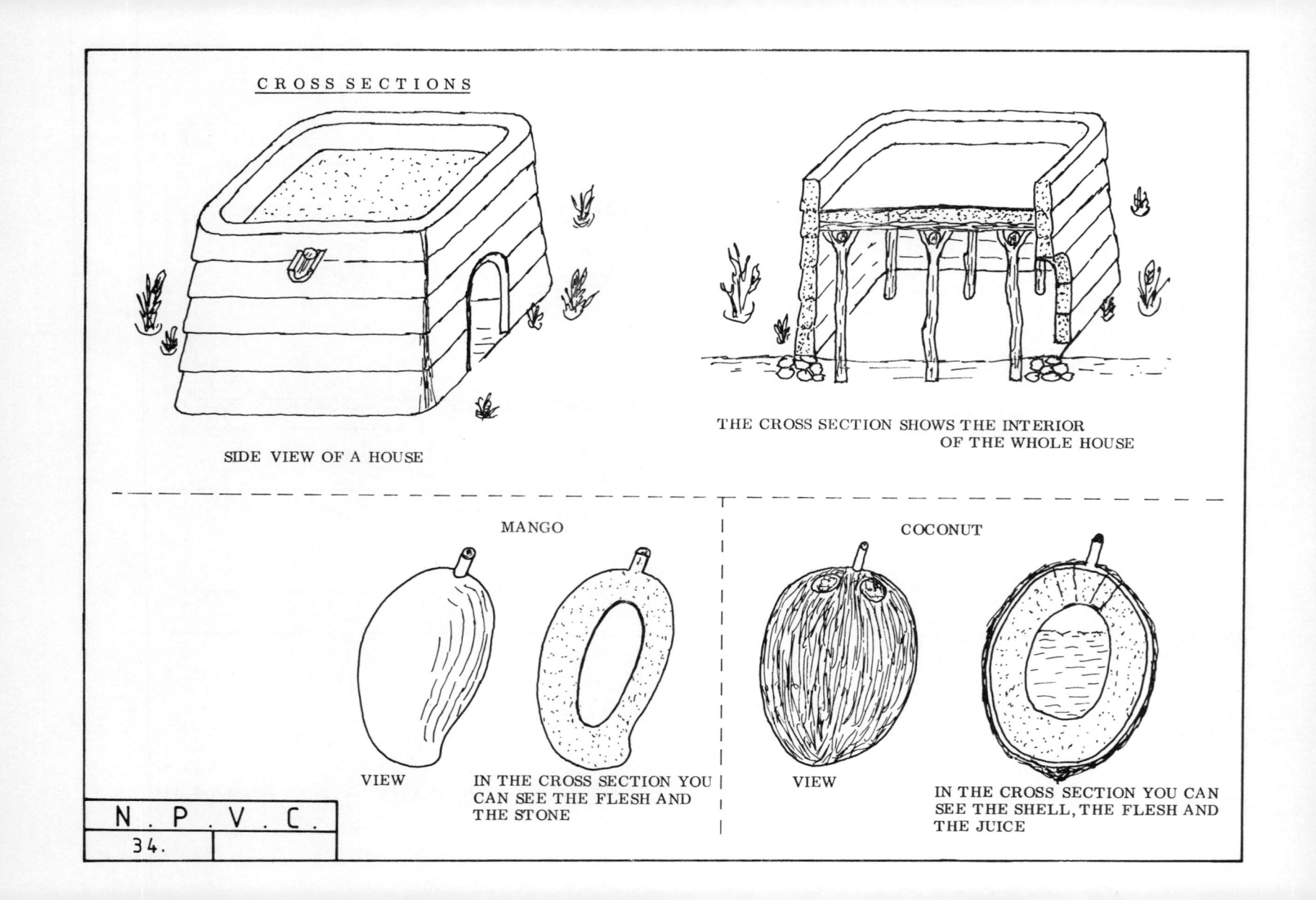

SIDE VIEW OF A HOUSE

THE CROSS SECTION SHOWS THE INTERIOR OF THE WHOLE HOUSE

MANGO

VIEW

IN THE CROSS SECTION YOU CAN SEE THE FLESH AND THE STONE

COCONUT

VIEW

IN THE CROSS SECTION YOU CAN SEE THE SHELL, THE FLESH AND THE JUICE

N.P.V.C.

HORIZONTAL SECTIONS AND CROSS SECTIONS

①

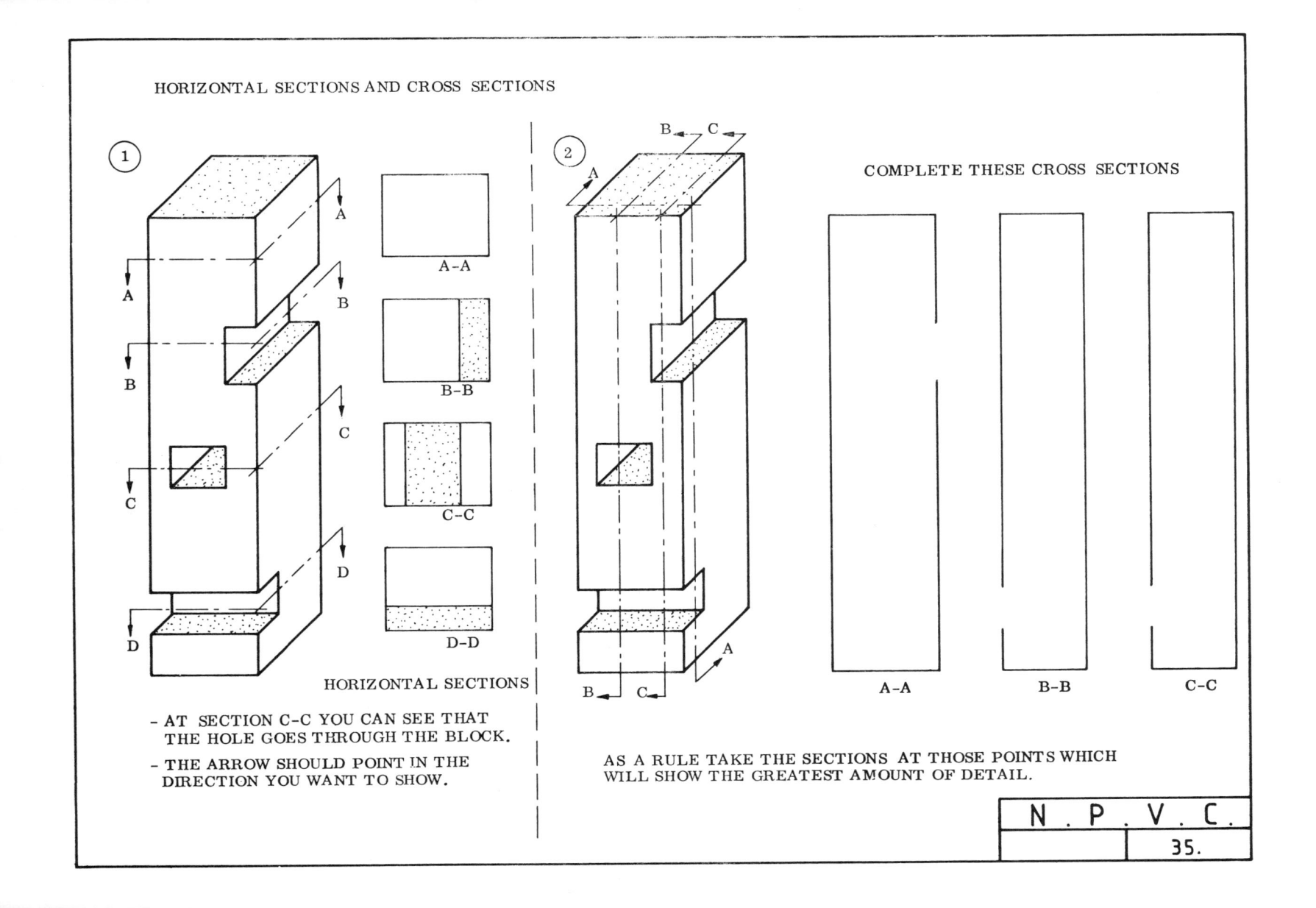

HORIZONTAL SECTIONS

- AT SECTION C-C YOU CAN SEE THAT THE HOLE GOES THROUGH THE BLOCK.
- THE ARROW SHOULD POINT IN THE DIRECTION YOU WANT TO SHOW.

②

COMPLETE THESE CROSS SECTIONS

A-A   B-B   C-C

AS A RULE TAKE THE SECTIONS AT THOSE POINTS WHICH WILL SHOW THE GREATEST AMOUNT OF DETAIL.

N. P. V. C.

35.

SKETCHING

You will often find that you will have to make a rough drawing in which accurate dimensions are not really necessary. In this case it is usually best to make a sketch. To sketch means to make a drawing without using drawing instruments like rulers, etc. You will find that this type of drawing is much quicker than technical drawing with the drawing instruments. However, sketching is probably harder than technical drawing, especially at first. It requires a steady hand, a sense for proportions, and an appreciation of detail.

Sketching is the art of putting ideas into pictures and is especially useful for understanding technical drawings and making rough plans.

The only equipment you need for sketching is an H pencil, an eraser, and some paper. Use plain paper for sketching.

- TECHNIQUE: Practice makes perfect. Practice drawing straight lines and curves. After a while you will find that your straight lines look almost as if they were drawn with a ruler! On the next page you can see the best way to draw horizontal, vertical, and inclined lines, and the best way to draw a circle.

- REMEMBER:

- Take your time while sketching, be careful.
- Don't try to draw long lines with one motion.
- Rest only the side of your hand on the paper.
- Don't turn the paper while you are sketching. You should be able to draw all your lines with the paper in the same position.

EXERCISE: FREE HAND SKETCHING

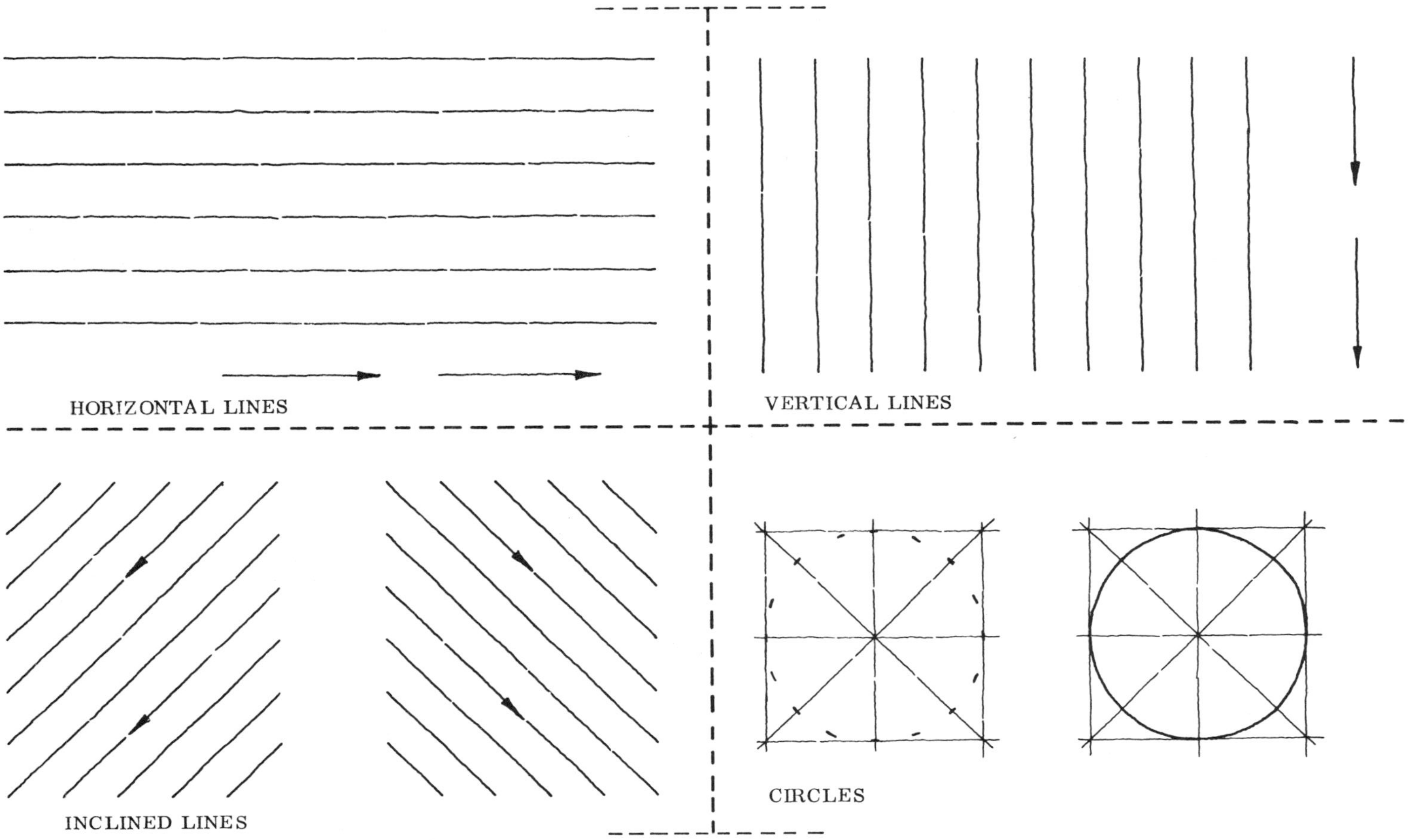

VERTICAL LINES

CIRCLES

HORIZONTAL LINES

INCLINED LINES

NOTICE: THE DIFFERENT DIRECTIONS IN WHICH THE LINES ARE DRAWN. LINES ARE NOT ONE LONG LINE.

N. P. V. C.

37.

EXERCISE: FREE HAND SKETCHING

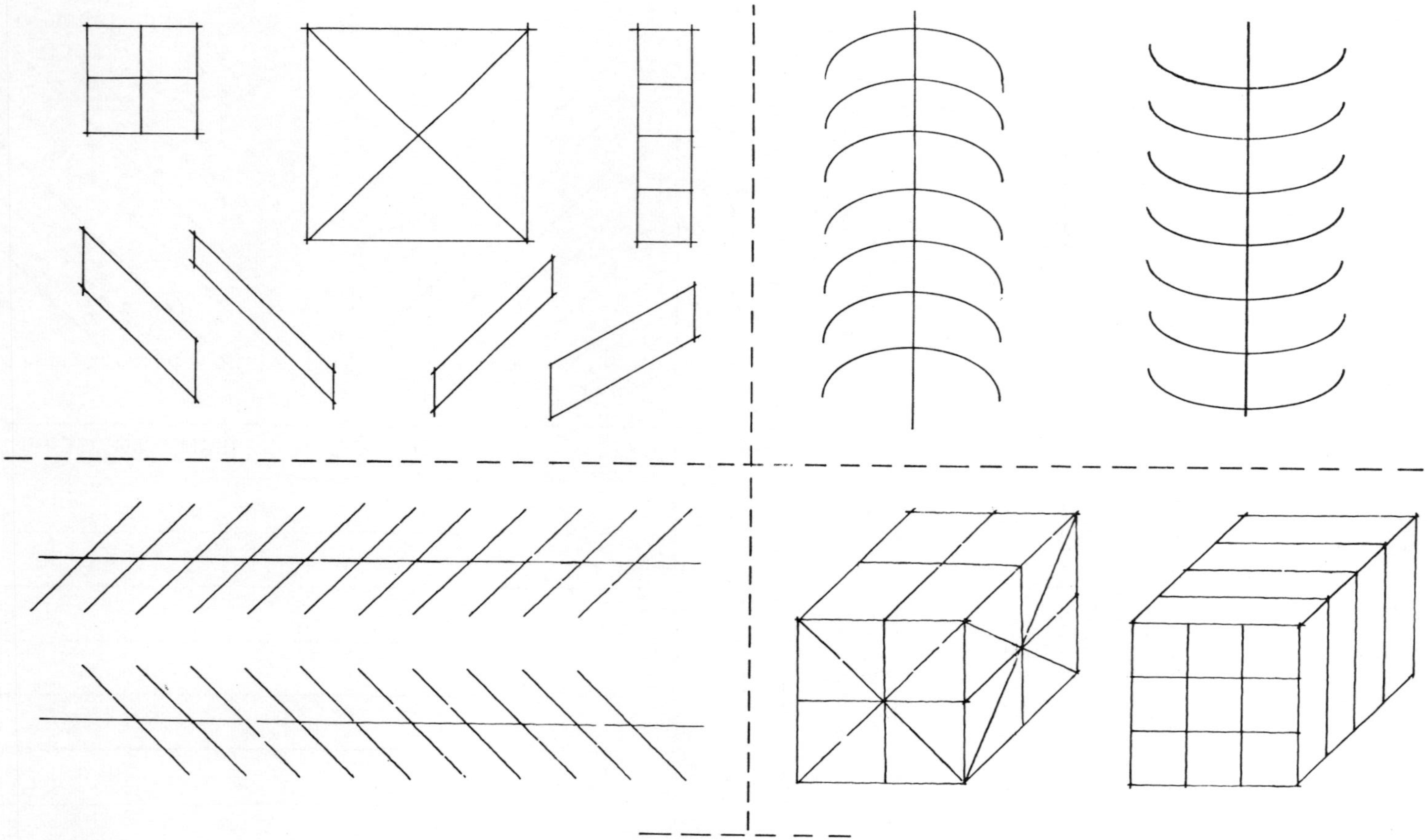

REPEAT THESE EXERCISES, USING DIFFERENT ANGLES.
DRAW ARROWS TO SHOW THE DIRECTION IN WHICH YOU SKETCH ALL OF THESE LINES.

N. P. V. C.

38.

EXERCISE: FREE HAND SKETCHING

A

90° 45° 60° 30° 60°

B

C

REPEAT THESE EXERCISES USING DIFFERENT ANGLES.

N. P. V. C.
39.

WINDOW FRAME (BASIC LAYOUT)

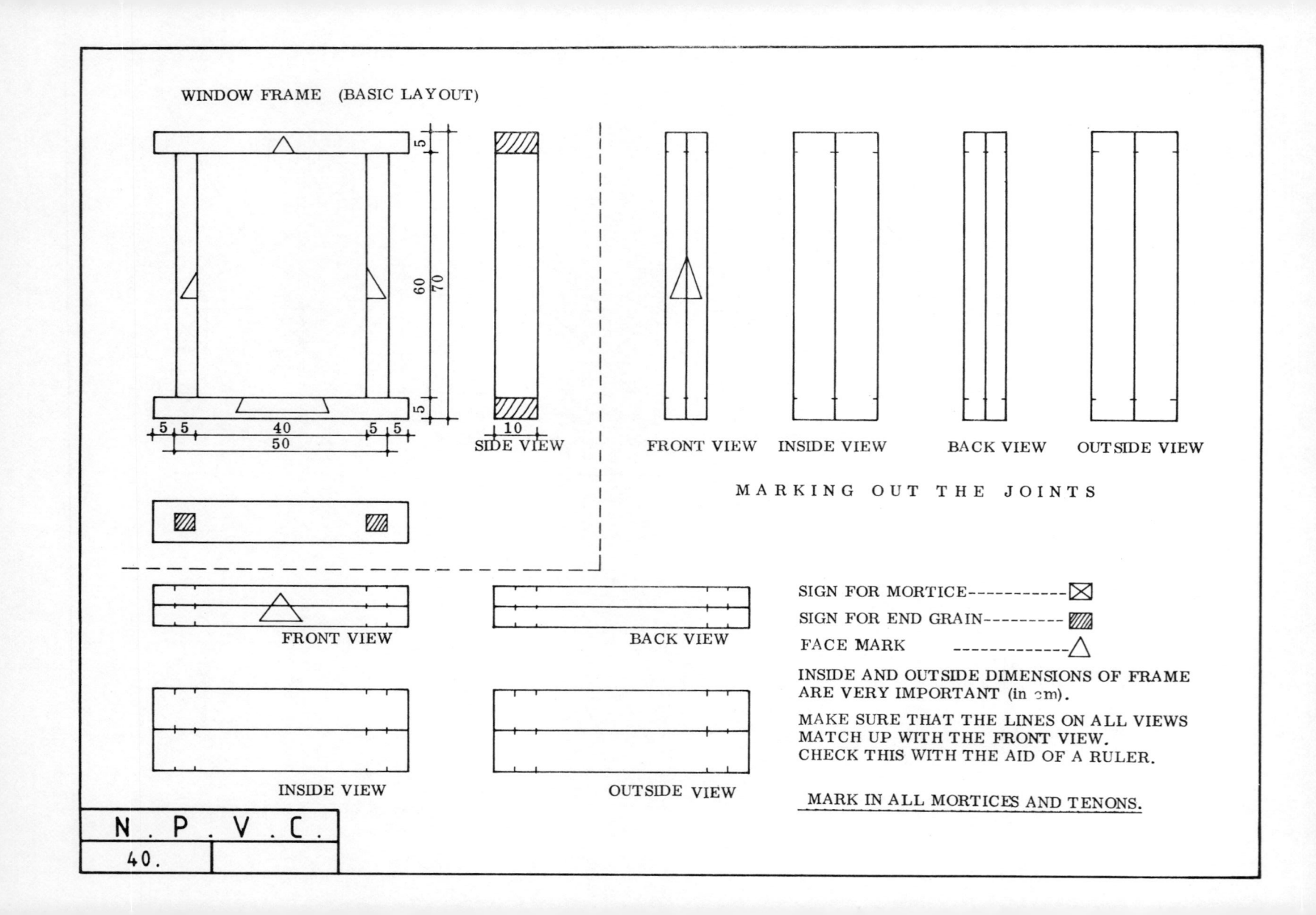

MARKING OUT THE JOINTS

FRONT VIEW   INSIDE VIEW   BACK VIEW   OUTSIDE VIEW

SIDE VIEW

FRONT VIEW   BACK VIEW

INSIDE VIEW   OUTSIDE VIEW

SIGN FOR MORTICE --------- ⊠
SIGN FOR END GRAIN ------- ▨
FACE MARK --------- △

INSIDE AND OUTSIDE DIMENSIONS OF FRAME ARE VERY IMPORTANT (in cm).

MAKE SURE THAT THE LINES ON ALL VIEWS MATCH UP WITH THE FRONT VIEW. CHECK THIS WITH THE AID OF A RULER.

MARK IN ALL MORTICES AND TENONS.

N.P.V.C.

40.

WINDOW FRAME (MARKED OUT FOR JOINTS)

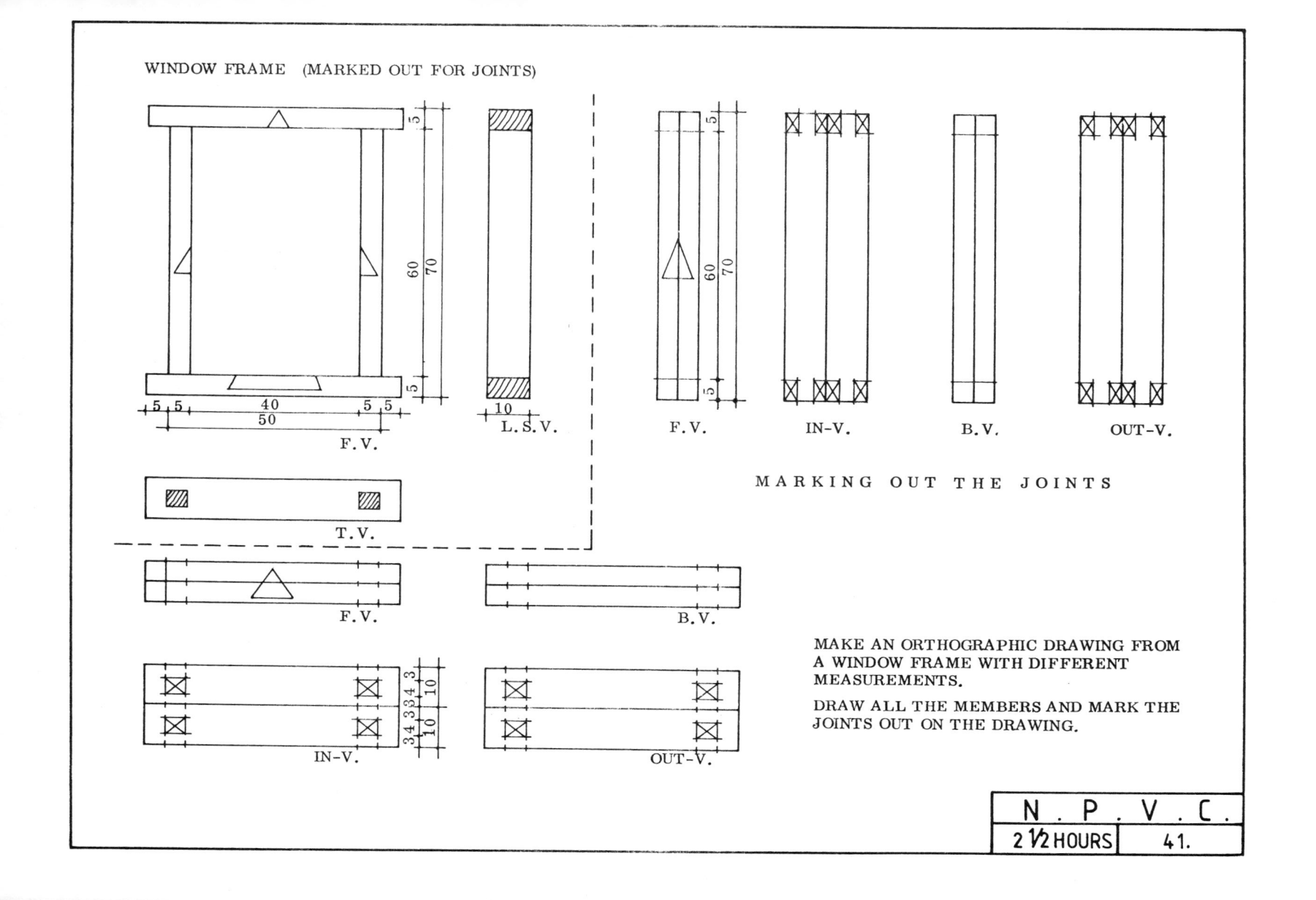

MARKING OUT THE JOINTS

MAKE AN ORTHOGRAPHIC DRAWING FROM A WINDOW FRAME WITH DIFFERENT MEASUREMENTS.

DRAW ALL THE MEMBERS AND MARK THE JOINTS OUT ON THE DRAWING.

N.P.V.C.
2½ HOURS | 4.1.

WINDOW FRAME   SCALE 1 : 10 (cm)

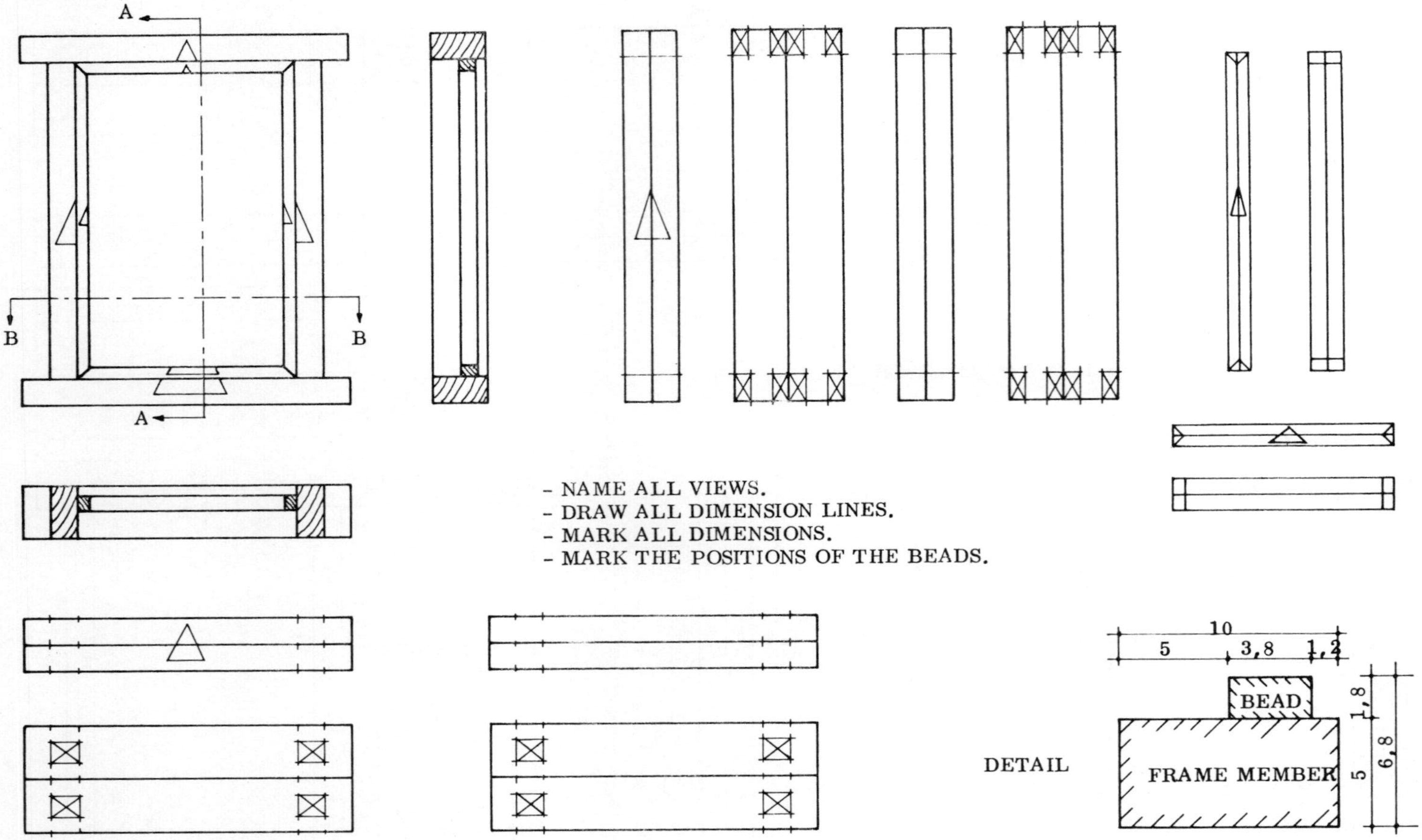

DETAIL

BEAD
FRAME MEMBER

- NAME ALL VIEWS.
- DRAW ALL DIMENSION LINES.
- MARK ALL DIMENSIONS.
- MARK THE POSITIONS OF THE BEADS.

MAKE A DRAWING OF A WINDOW FRAME WITH DIFFERENT DIMENSIONS.

N. P. V. C.
42.   2 ½ HOURS

DOOR FRAME FOR A SMALL DOOR   SCALE 1 : 10 (cm)

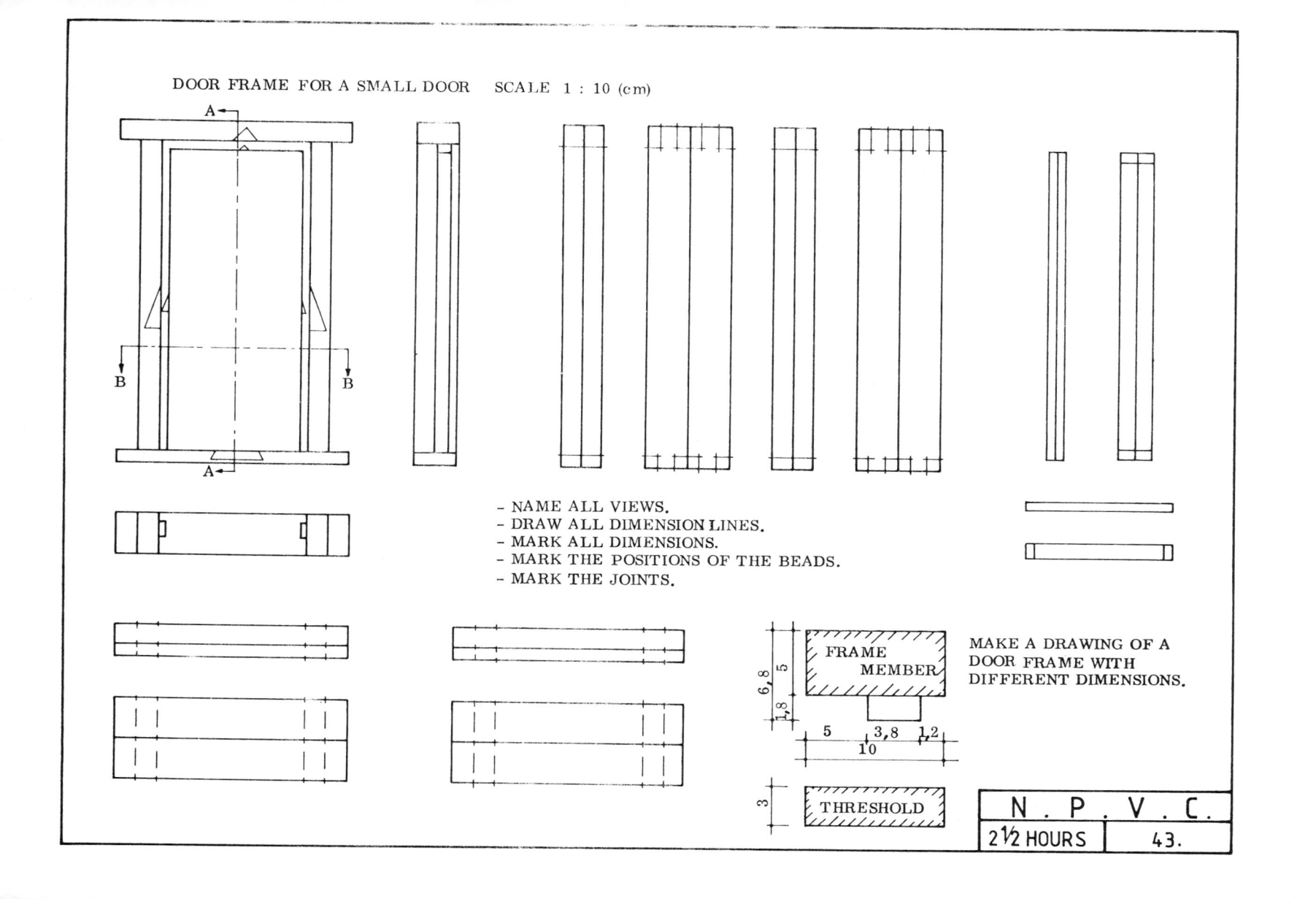

FRAME MEMBER
6,8
5   1,8
5   3,8 1,2
10

THRESHOLD
3

MAKE A DRAWING OF A
DOOR FRAME WITH
DIFFERENT DIMENSIONS.

- NAME ALL VIEWS.
- DRAW ALL DIMENSION LINES.
- MARK ALL DIMENSIONS.
- MARK THE POSITIONS OF THE BEADS.
- MARK THE JOINTS.

N. P. V. C.

2½ HOURS   43.

WOODEN FRAME SCALE 1 : 10 (cm)

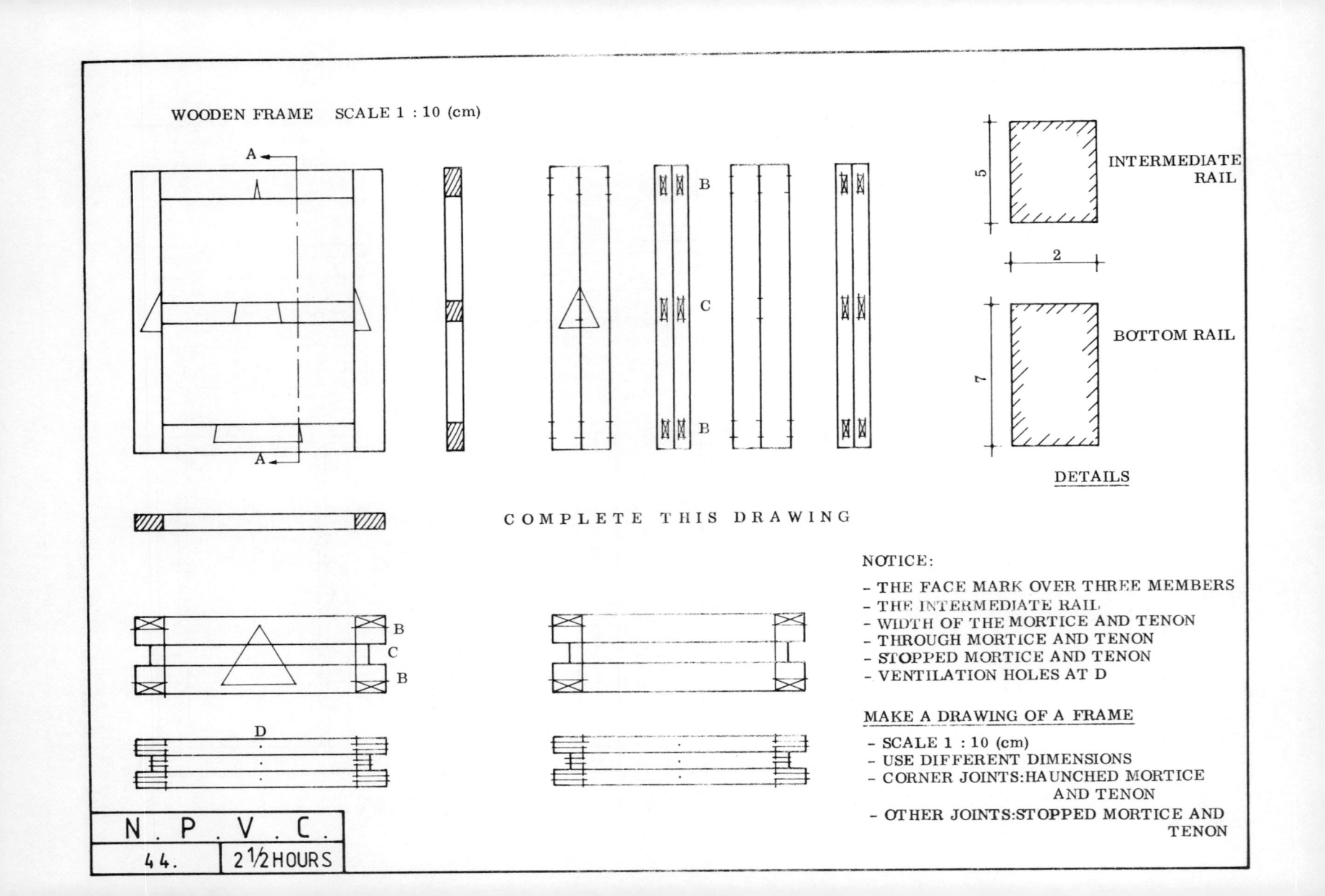

DETAILS

INTERMEDIATE RAIL — 5 × 2

BOTTOM RAIL — 7 × 2

COMPLETE THIS DRAWING

NOTICE:
- THE FACE MARK OVER THREE MEMBERS
- THE INTERMEDIATE RAIL
- WIDTH OF THE MORTICE AND TENON
- THROUGH MORTICE AND TENON
- **STOPPED** MORTICE AND TENON
- VENTILATION HOLES AT D

MAKE A DRAWING OF A FRAME
- SCALE 1 : 10 (cm)
- USE DIFFERENT DIMENSIONS
- CORNER JOINTS: HAUNCHED MORTICE AND TENON
- OTHER JOINTS: STOPPED MORTICE AND TENON

N. P. V. C.

4.4.   2½ HOURS

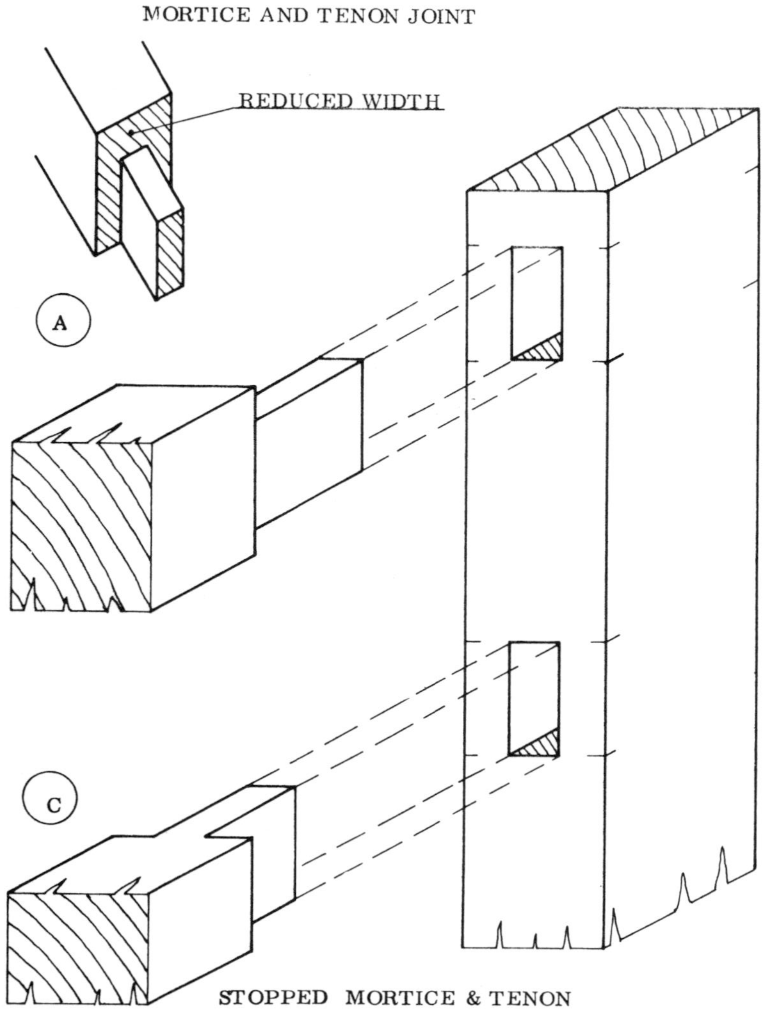

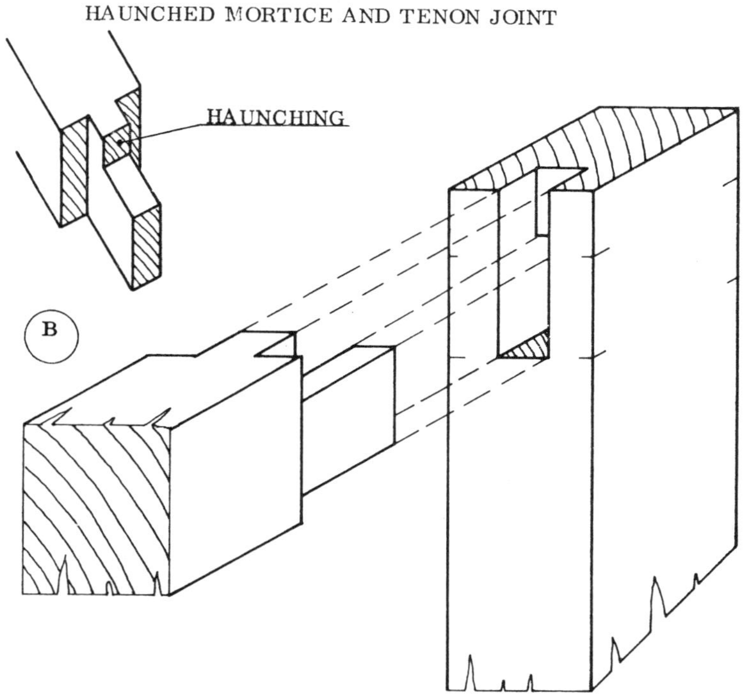

MORTICE AND TENON JOINT

A — REDUCED WIDTH

STOPPED MORTICE & TENON

C

HAUNCHED MORTICE AND TENON JOINT

B — HAUNCHING

- MAKE AN OBLIQUE DRAWING OF:

  A STOPPED HAUNCHED MORTICE AND TENON JOINT.
  USE DIFFERENT POSITIONS.

N. P. V. C.

45.

WOODEN FRAME WITH SLANTING CILL

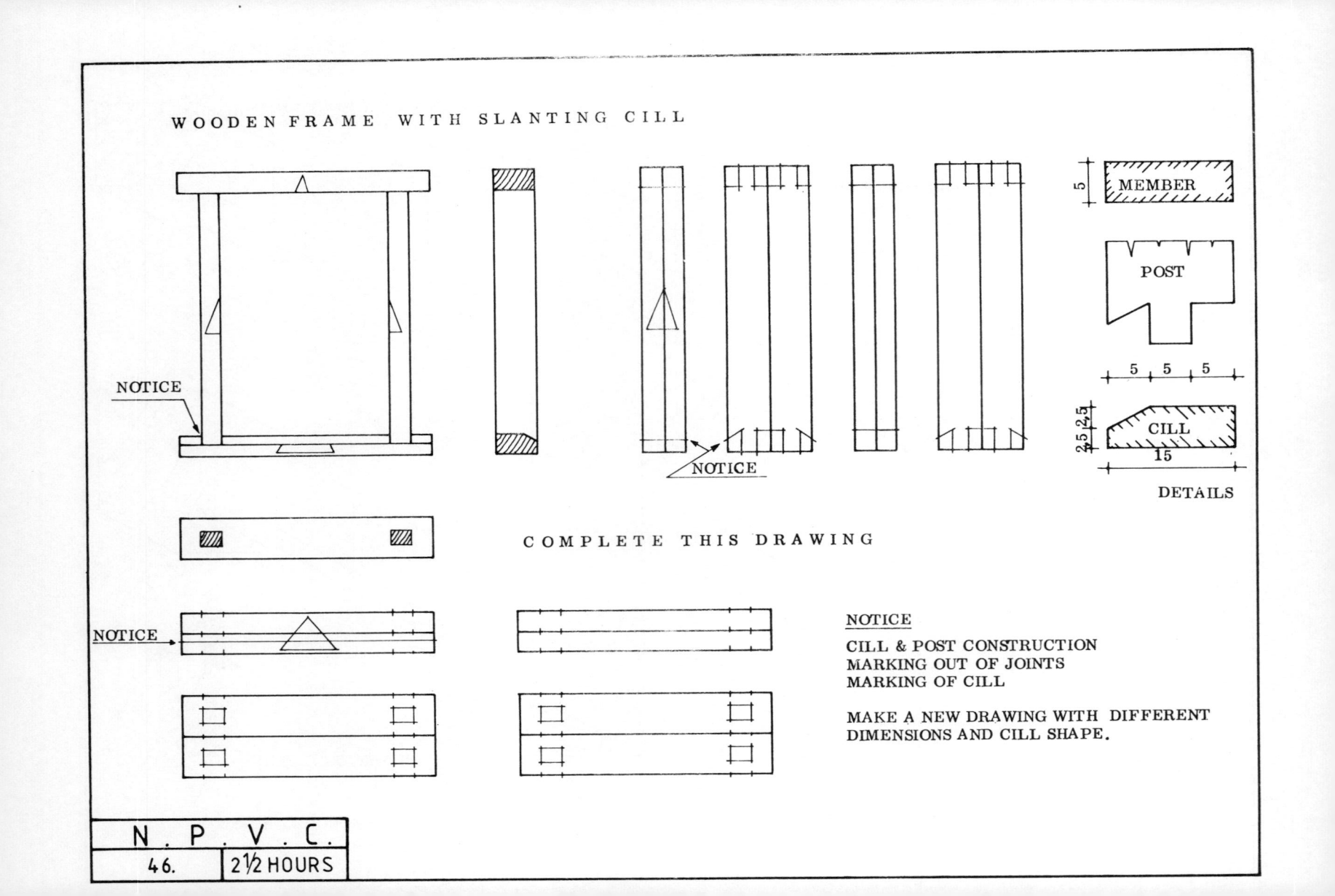

MEMBER
POST
CILL
DETAILS

COMPLETE THIS DRAWING

NOTICE

CILL & POST CONSTRUCTION
MARKING OUT OF JOINTS
MARKING OF CILL

MAKE A NEW DRAWING WITH DIFFERENT
DIMENSIONS AND CILL SHAPE.

N. P. V. C.
46. 2½ HOURS

OBLIQUE DRAWING OF "CILL-POST" CONSTRUCTION

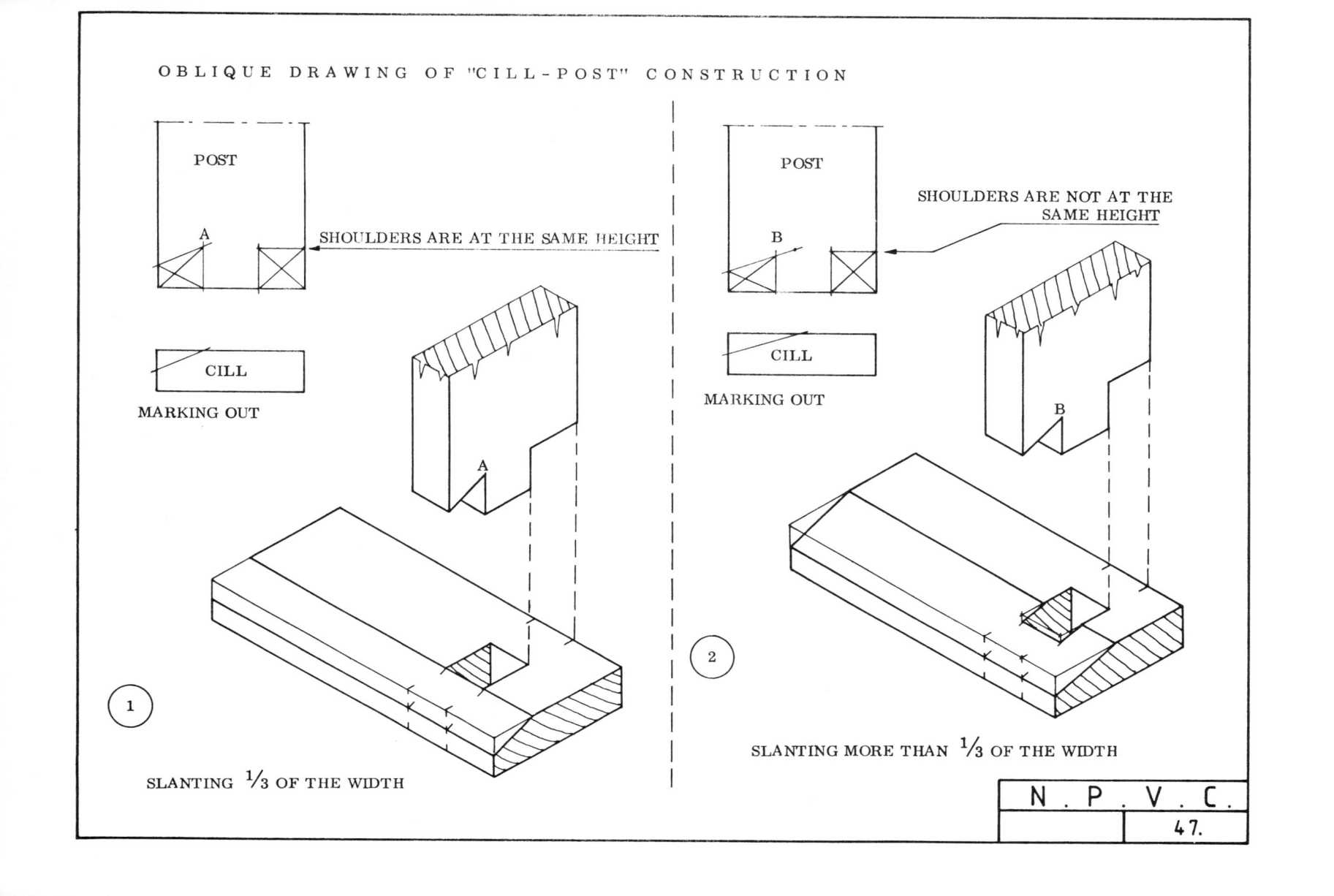

1

SHOULDERS ARE AT THE SAME HEIGHT

POST

CILL

MARKING OUT

SLANTING 1/3 OF THE WIDTH

2

SHOULDERS ARE NOT AT THE SAME HEIGHT

POST

CILL

MARKING OUT

SLANTING MORE THAN 1/3 OF THE WIDTH

N.P.V.C.

REMEMBER

NOW THIS BOOK IS REALLY YOURS !
FOR YOU IT OPENS MANY DOORS
TO KNOWLEDGE. ON ITS PAGES WRITE
NOTES TO HELP YOU UNDERSTAND, MAKE LIGHT
SKETCHES AND DRAWINGS ALL YOUR OWN.

THOUGHTS OF YOURS, LIKE SEED ARE SOWN
FRUITS COME LATER UNDER SUN AND RAIN
IN THE FORM OF BUILDINGS WITH DOORS, ROOF AND WINDOW-PANE.

KEEP YOUR WORK BOLD, CLEAR AND CLEAN
THIS BOOK WILL BE YOUR PRIDE, AND GUIDE TO ALL
YOU'VE LEARNT AND SEEN.

FORM 2: GENERAL BUILDING INFORMATION

The important rules for a Rural Builder to remember in general are:

- PLAN CAREFULLY EVERYTHING THAT YOU MAKE OR BUILD.
- DRAW ALL IDEAS OUT ON PAPER FOR THE BENEFIT OF THOSE WHO WILL HAVE A PART IN THE PROJECT BEFORE, DURING AND AFTER CONSTRUCTION.
- ALWAYS TAKE INTO ACCOUNT MATERIALS WHICH ARE LOCALLY AVAILABLE, AND WHEN POSSIBLE USE THESE RATHER THAN EXPENSIVE IMPORTED MATERIALS.
- TAKE INTO ACCOUNT THE SIZES OF READY-MADE MATERIALS WHEN YOU ARE PLANNING THE BUILDING.
- BE AWARE OF THE REQUIREMENTS OF THE ENVIRONMENT AND THE LOCAL CUSTOMS.

Drawings should be precise and clear and should take account of the sizes of all the ready-made materials, so as to avoid unnecessary waste. It is therefore important to be informed about all the materials which are available for Rural Building, and their sizes. These can be found in the Reference Book, Materials and Products sections.

Remember that a building is an investment and the construction should be long-lasting. Proper construction, using good materials, will avoid unnecessary expense, inefficiency, and dissatisfaction.

As a rule, mistakes in design or construction are costly, obvious, and permanent.

BUILDING KEY

BUILDING KEY

1 = GROUND LEVEL
2 = REINFORCED CONCRETE
3 = CONCRETE
4 = ONE COURSE WORK
5 = TWO COURSE WORK
6 = HARD CORE FILLING
7 = WOOD (large)
8 = WOOD (small)
9 = FINISHED FLOOR LEVEL
10 = PLASTER
11 = LANDCRETE WALL
12 = SANDCRETE WALL
13 = SANDCRETE TO LAND-CRETE

N.P.V.C.

49.

BUILDING KEY

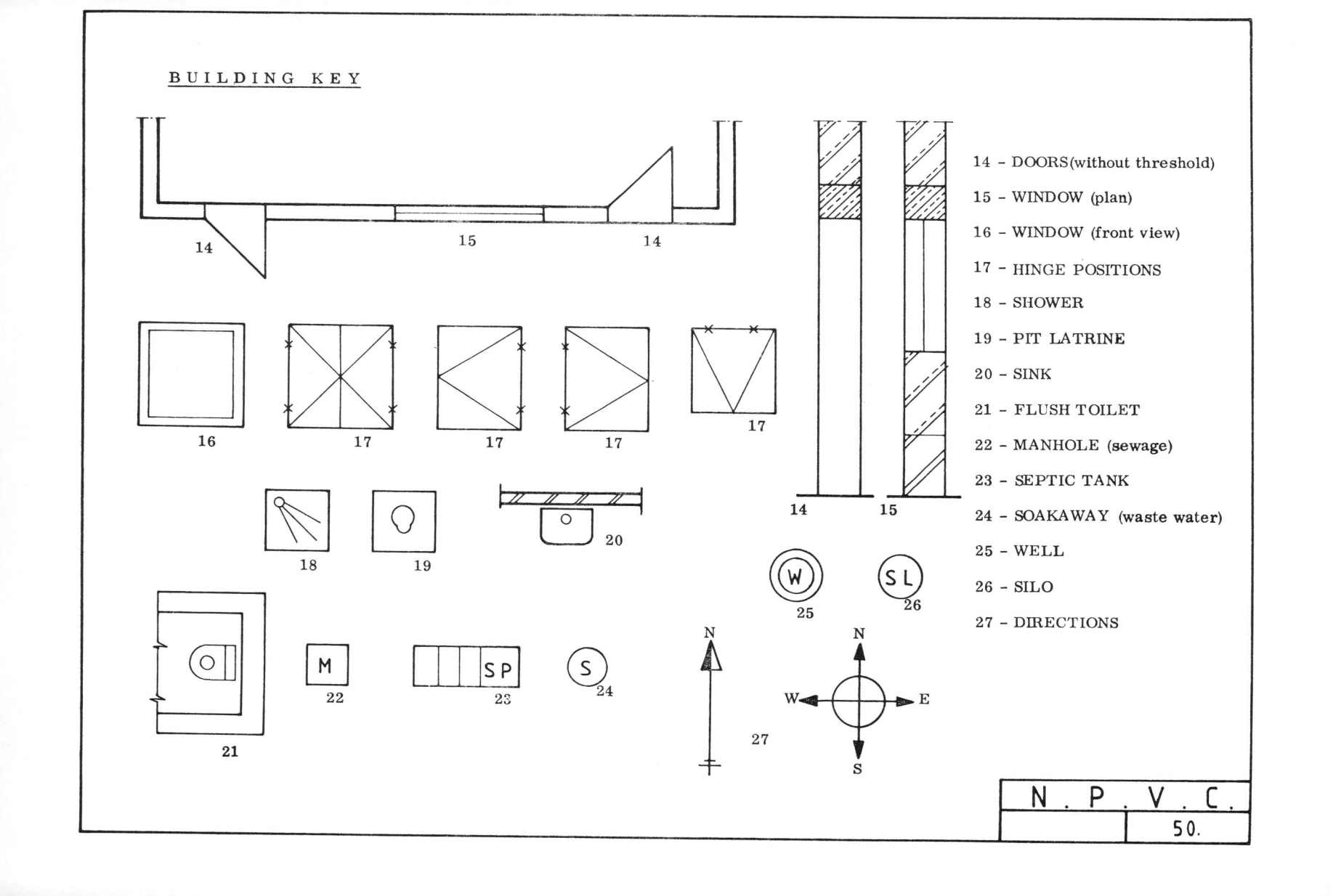

14 – DOORS (without threshold)
15 – WINDOW (plan)
16 – WINDOW (front view)
17 – HINGE POSITIONS
18 – SHOWER
19 – PIT LATRINE
20 – SINK
21 – FLUSH TOILET
22 – MANHOLE (sewage)
23 – SEPTIC TANK
24 – SOAKAWAY (waste water)
25 – WELL
26 – SILO
27 – DIRECTIONS

HOW BUILDINGS ARE DRAWN

SCALE

When you draw a plan of a building on paper you will find that you have to use a scale, to make the plan small enough to fit on the paper. Scales were explained earlier, but here are some examples to help you remember how to use them.

- EXAMPLES: A scale of 1:50 (cm) tells you that 1 cm on the drawing represents 50 cm in real life. In other words all the dimensions on the drawing are 1/50th of their real size and all the dimensions are in centimetres.

A scale of 1:500 (cm) tells you that all the dimensions on the drawing are 1/500th of their real size and that all the dimensions are in centimetres.

Both of these scales make the drawing smaller than the actual building size. These types of scales are called "reduced scales".

The man you see on the next page is making a scale drawing of an electricity pole. The pole is 800 cm high and the man is using a scale of 1:100 (cm) which means that his drawing will be 8 cm high. The 8 cm on the drawing represents 800 cm in real life.

If the crossbar on the pole is 150 cm long, how long will the man draw it on his paper?

All building drawings are made in reduced scale. Here are some examples of drawings used in building, and the scales commonly used with them:

- LOCATION PLAN   1 : 500
- DESIGN DRAWING   1 : 100
- FINAL DRAWING   1 : 50
- DETAIL DRAWING   1 : 20; 1 : 10, or 1 : 5

Always remember to include the UNITS of the scale you have used; these are usually "cm" or "mm".

IF THIS POLE IS 8 METER HIGH

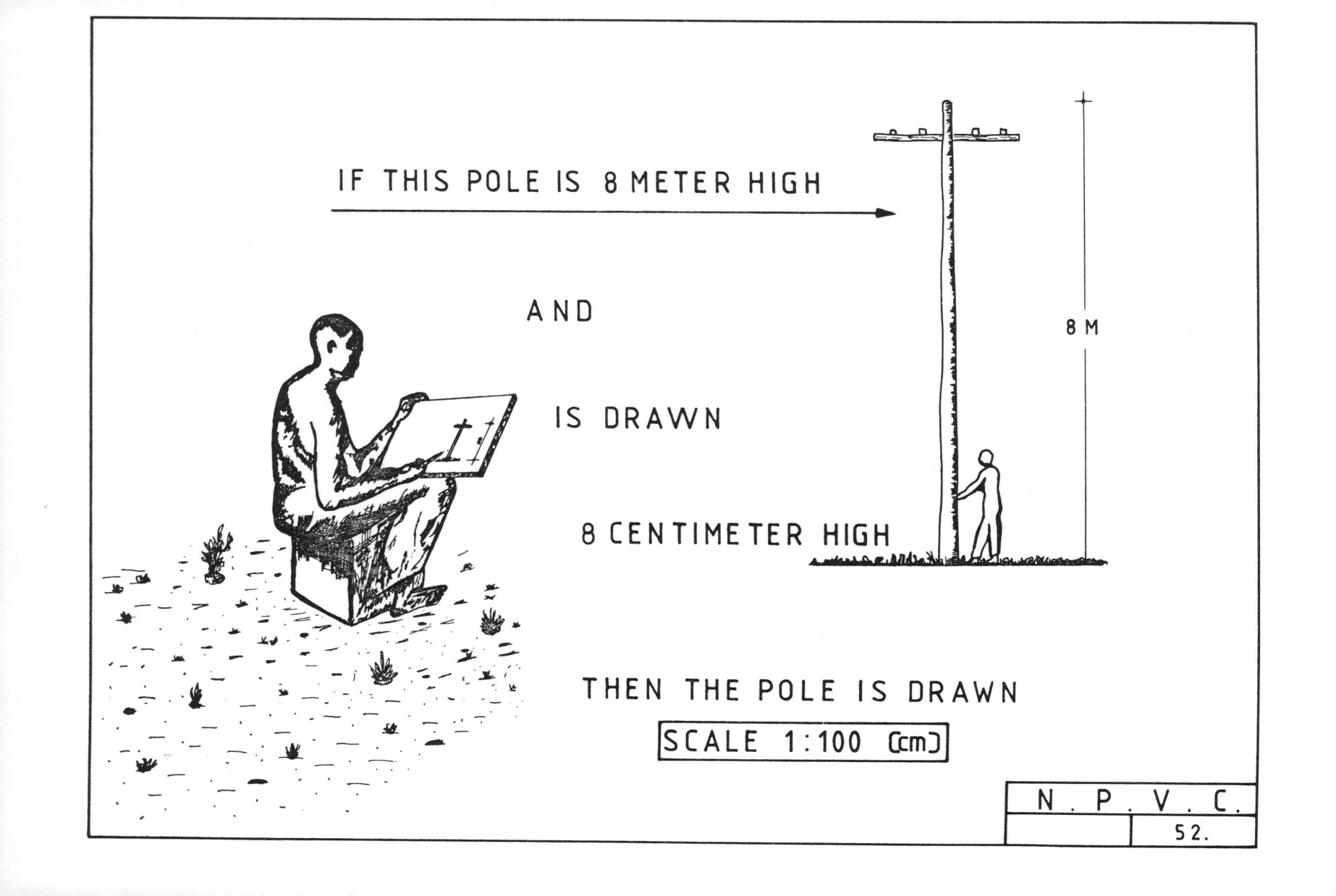

AND

IS DRAWN

8 CENTIMETER HIGH

THEN THE POLE IS DRAWN

8 M

SCALE 1:100 (cm)

N.P.V.C.

52.

WORKING DRAWINGS

Three different types of drawings are needed to show the builder exactly how the building should look, on the inside and the outside. These include the elevations and sections, as well as the plans. Here we describe the different types of plans which have to be made: the foundation plan, floor plan, and roof plan.

- FOUNDATION PLAN: A foundation plan, along with its sections, shows the builder how deep the foundations should be laid and gives all the dimensions for the foundation and the footings. Sometimes the corners of the rising walls are indicated on the footings.

- FLOOR PLAN: This plan should show the builder the size of the building and the verandahs, the thickness of the walls, and where to place the doors and windows. It also shows which way the doors are meant to open.

- ROOF PLAN: Roof plans are made to show the builder what shape the roof should be and how it is to be built. The roof plan should contain such information as the angle of the roof, the shape, and the materials to be used.

The drawing here illustrates the types of plans and what it is they show to the builder.

Elevations and cross sections are of course essential parts of the working drawings. These are examined in detail after some further examples of plans in the next pages.

PLANS

ROOF — DRAWING → ROOF PLAN

RISING WALLS — DRAWING → FLOOR PLAN

FOOTINGS & FOUNDATIONS — DRAWING → FOUNDATION PLAN

N.P.V.C.
54.

FLOOR PLANS

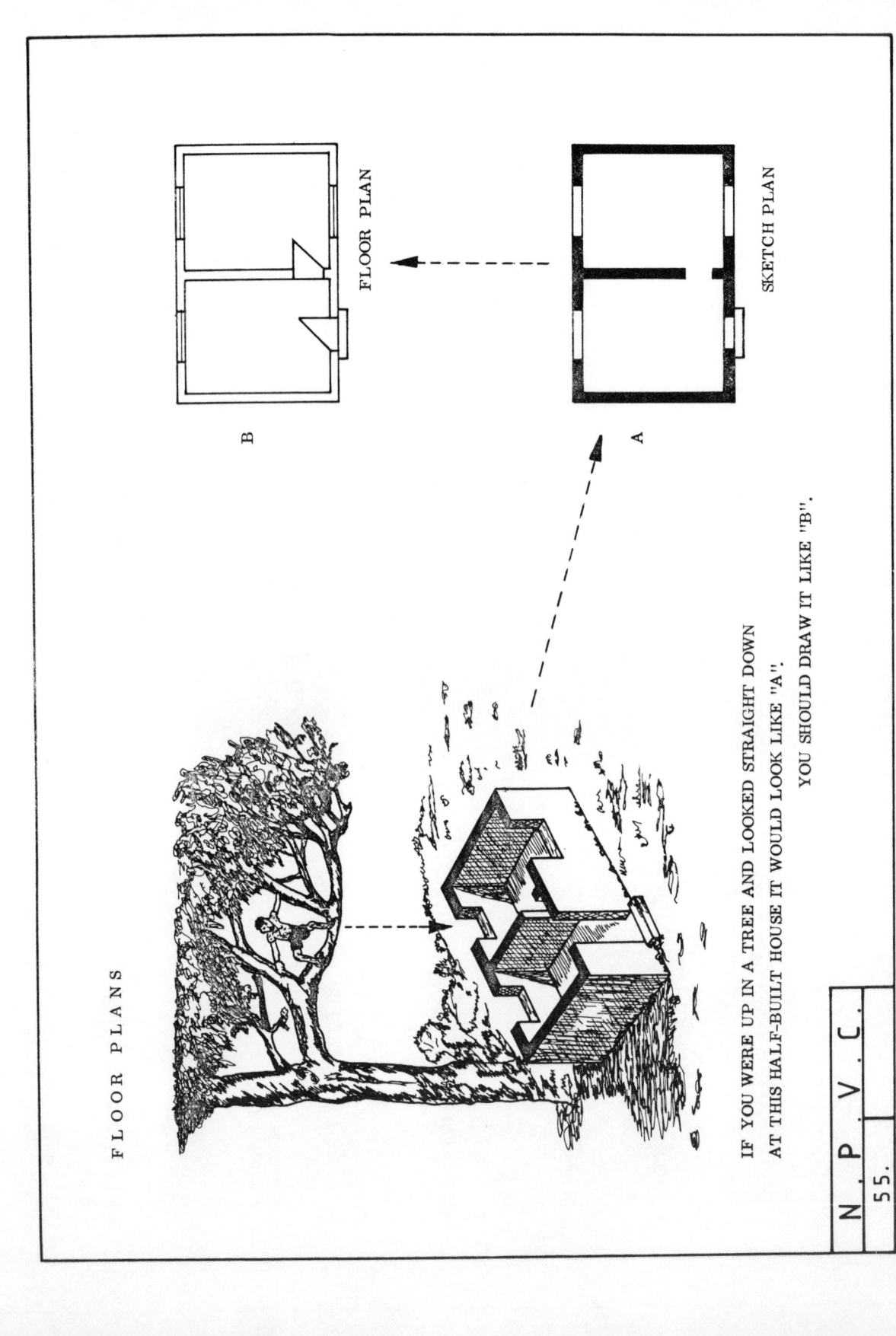

FLOOR PLAN

SKETCH PLAN

B

A

IF YOU WERE UP IN A TREE AND LOOKED STRAIGHT DOWN AT THIS HALF-BUILT HOUSE IT WOULD LOOK LIKE "A".
YOU SHOULD DRAW IT LIKE "B".

N. P. V. C.
55.

FLOOR PLAN

DIMENSION LINE A = OPENINGS
,,   ,,   B = WALLS
,,   ,,   C = TOTAL LENGTH

DIMENSIONS IN CM

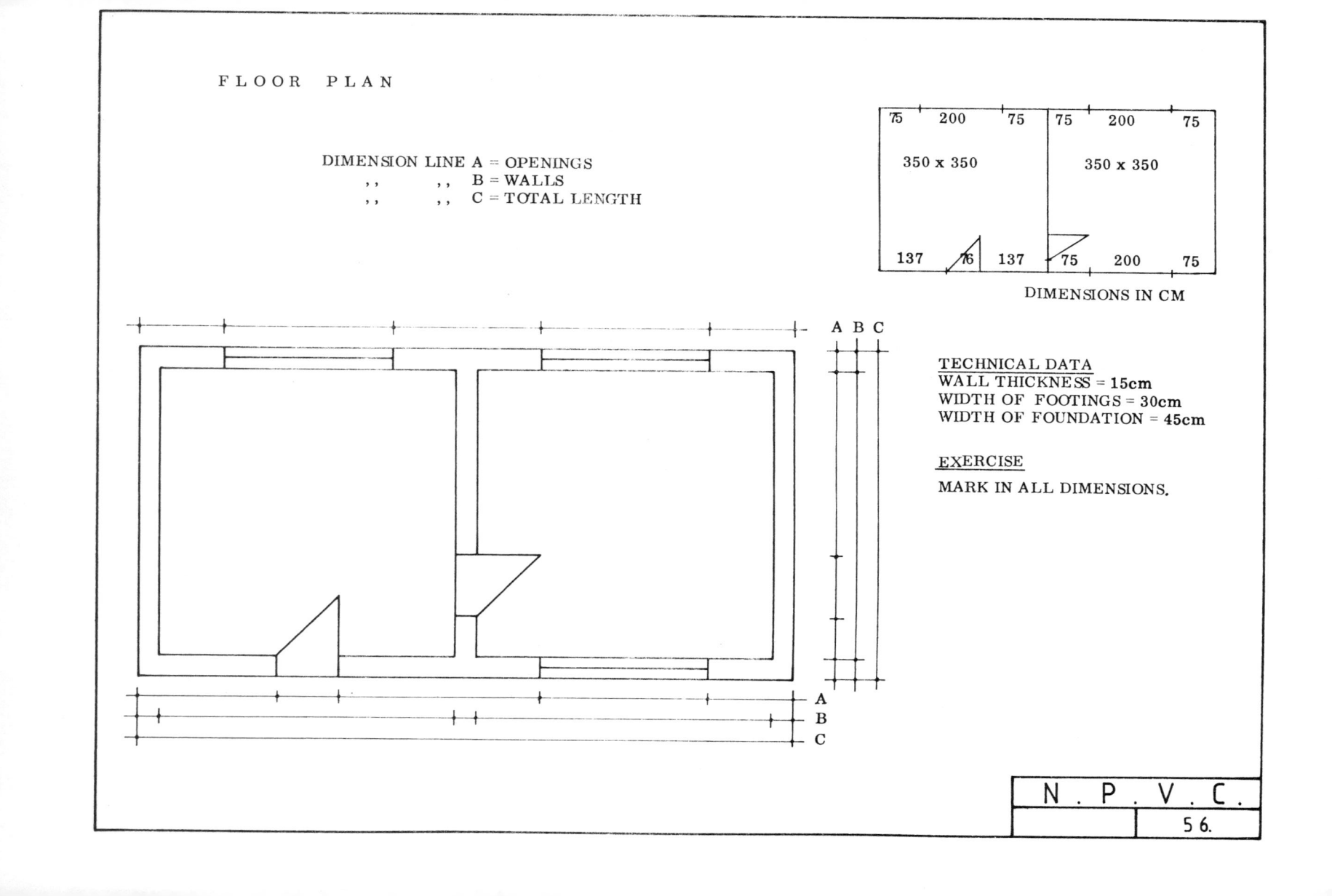

TECHNICAL DATA
WALL THICKNESS = 15cm
WIDTH OF FOOTINGS = 30cm
WIDTH OF FOUNDATION = 45cm

EXERCISE
MARK IN ALL DIMENSIONS.

N.P.V.C.
56.

FLOOR PLAN

WINDOW AND DOOR HEIGHTS

SKETCH PLAN

PLANNING LINE A

PLANNING LINE B

MEASUREMENTS OF AREAS ARE ALWAYS INSIDE MEASUREMENTS!

FFL
TOTAL HEIGHT

EXERCISE
- MARK IN ALL DIMENSIONS ON PLAN.
- MARK IN ALL DIMENSIONS FOR DOOR AND WINDOW HEIGHTS.
- MARK IN PLANNING LINES A & B.
- FIND THE TOTAL LENGTH AND WIDTH (SCALE = 1 : 100 cm).

N.P.V.C.

57.

SKETCHES FOR FLOOR PLANS

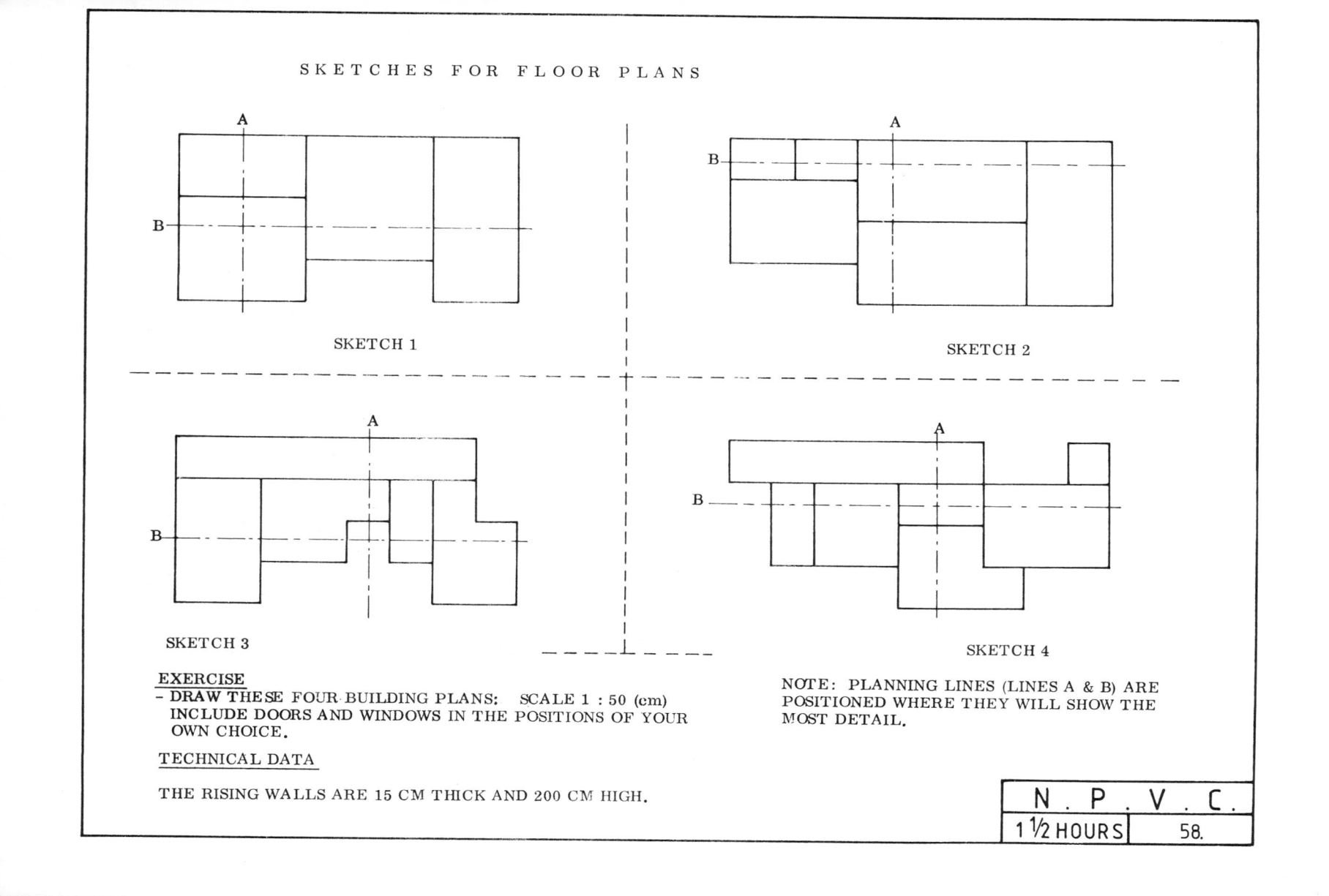

SKETCH 1

SKETCH 2

SKETCH 3

SKETCH 4

NOTE: PLANNING LINES (LINES A & B) ARE POSITIONED WHERE THEY WILL SHOW THE MOST DETAIL.

EXERCISE
- **DRAW THESE FOUR BUILDING PLANS; SCALE 1 : 50 (cm)**
- **INCLUDE DOORS AND WINDOWS IN THE POSITIONS OF YOUR OWN CHOICE.**

TECHNICAL DATA

THE RISING WALLS ARE 15 CM THICK AND 200 CM HIGH.

N. P. V. C.
1½ HOURS   58.

ELEVATIONS

A special type of drawing is used to show what a building will look like from the outside. These drawings are called "elevations" and they show what you would see if you looked straight on at the side of the house. Of course a house has more than one side, and so there are always a number of elevations. There are as many elevations as there are sides of the house. Houses usually have four sides and so there are usually four elevations.

It is not always necessary to draw all the possible elevations of a building, especially if some of the sides look very similar to each other.

The drawing here shows the front elevation of a small house. You should notice that the sides of the house are not drawn. This is because if you stand directly in front of the house you cannot see the sides. The front elevation shows the sizes and positions of the doors and windows as well as the height and length of the house itself.

The building shown on the next page faces south. The front elevation is therefore called the "south elevation". The other three elevations are the east elevation, the north elevation, and the west elevation.

In general only main features are shown on the elevations. Small details of the doors, windows, etc. are given in the detail drawings.

ELEVATIONS

ELEVATION "A"

IF YOU LOOKED STRAIGHT AT ONE SIDE OF
A BUILDING IT WOULD LOOK LIKE "A".

YOU SHOULD DRAW IT LIKE "B".

FRONT ELEVATION "B"

FRONT ELEVATION "A"

N.P.V.C.

60.

SECTIONS

Suppose you were able to cut right through a building and then take away one part. What you would see would look something like the diagram on the next page (1). If you now look at the building straight on, you will see a "cross section" of the building (2). The cross section shows the insides of the roof and the room as well as the footings and foundations.

Sections are useful because they give a lot of information about the building which is not found in the elevations and the plans. For example, on a section you can see the height of a room inside the building, the thickness of the ceiling, and the floor and roof construction. You can also see the thickness and width of the foundation, which is not given in the plans and elevations.

- CHOOSING SECTIONS: You will usually find it necessary to take at least two sections through the building. You can take any section through any part of the building, but of course the best sections are the ones which are the hardest to draw! When you take a section through a building, you have to mark on the plan exactly where you have "cut" and the direction from which you look at the section.

On the right page you can see the conventional way of doing this. The place where the building is cut is marked by a broken line which has arrows at its ends to show which way the section faces. All sections should be marked on the plan, and you should label each end of the line with a letter. On the plan here, the section has the letter "A" at each end. When this cross section is drawn, it is labelled as "cross section A-A". The next section would be "cross section B-B" etc.

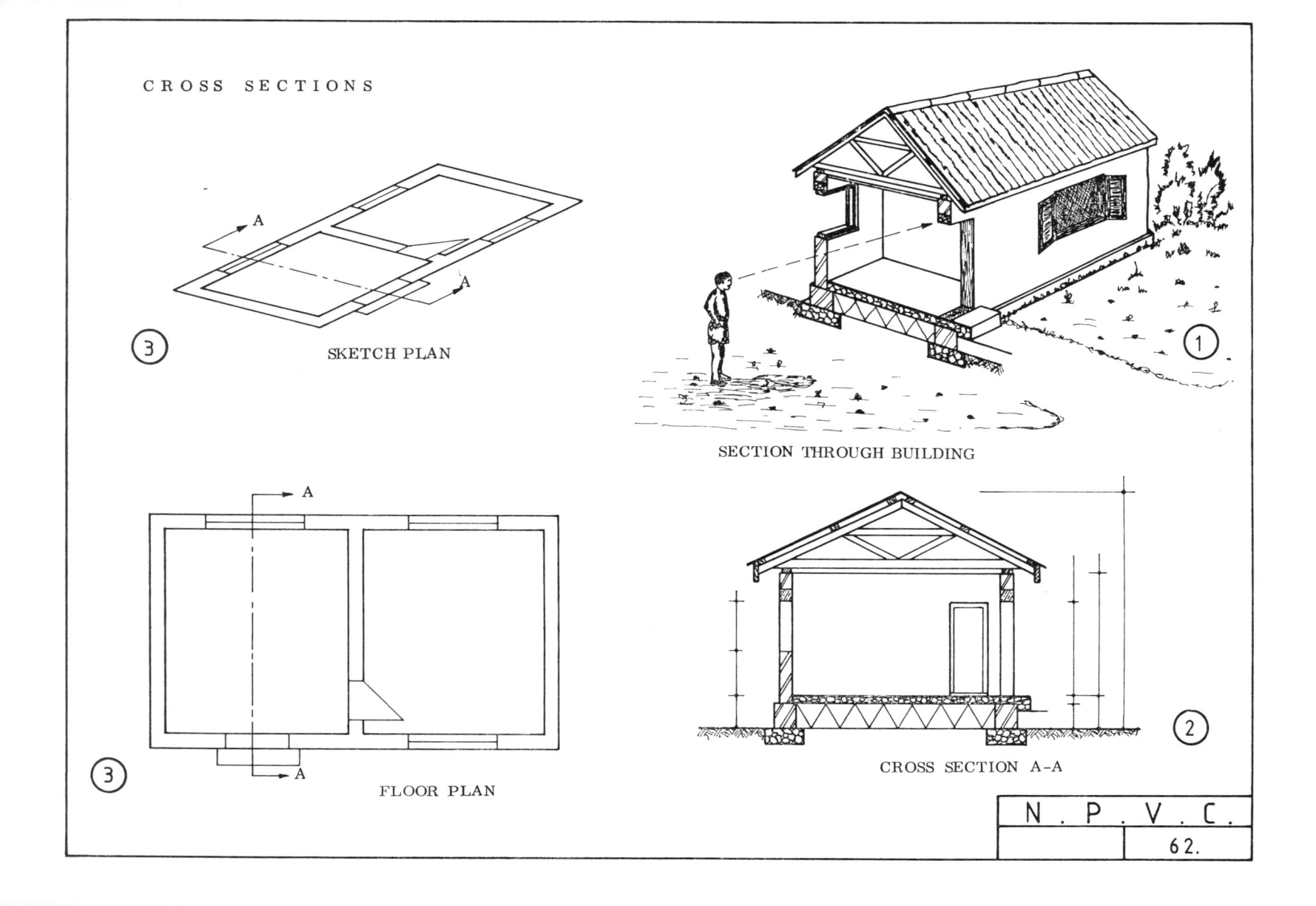

SECTION THROUGH BUILDING

CROSS SECTION A-A

CROSS SECTIONS

SKETCH PLAN

FLOOR PLAN

N. P. V. C.

62.

PLANS - ELEVATIONS - CROSS SECTIONS

RIGHT SIDE VIEW

LEFT SIDE VIEW

CROSS SECTION A-A

BACK VIEW

FRONT VIEW

PLAN

IMPORTANT
- INDICATE ALL DIMENSIONS AT LEAST ONCE.
- MAKE CROSS SECTIONS OF THE MOST DIFFICULT PARTS OF THE BUILDING.
- ALL ELEVATIONS SHOULD BE IN LINE WITH EACH OTHER.
- IF THERE IS NOT ENOUGH SPACE MARK DIMENSIONS LIKE AT "D".

NOTICE
- THE FLOOR GOES THROUGH AT "A".
- THE FLOOR IS STOPPED AT "B".
- FLOOR THICKNESS AT "C".

EXERCISE
MAKE A SIMILIAR DRAWING WITH DIFFERENT DIMENSIONS AND DIFFERENT DESIGN.

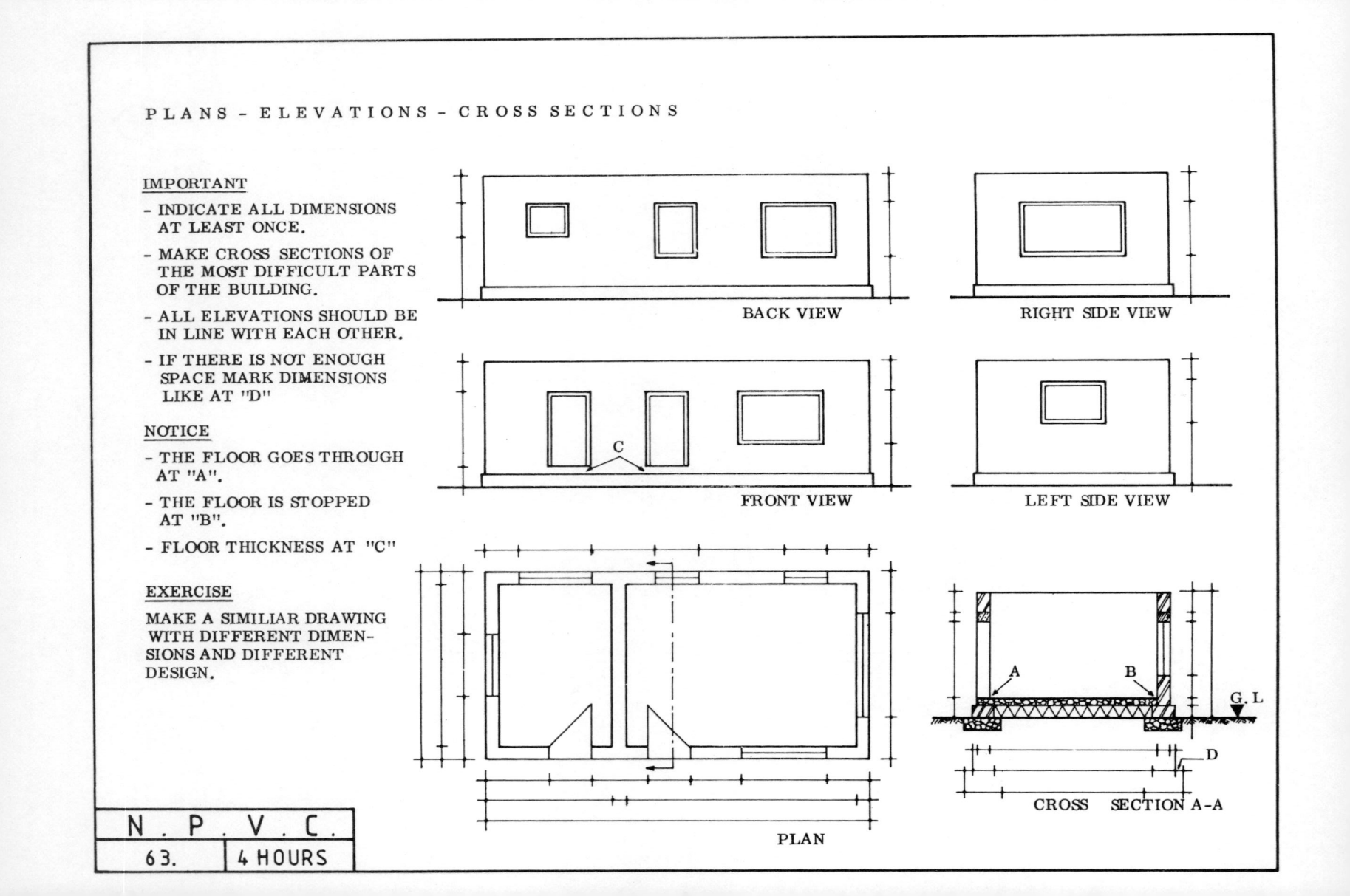

N. P. V. C.

63. | 4 HOURS

FOUNDATION PLAN

MAKE THE PLANNING LINES FIRST!

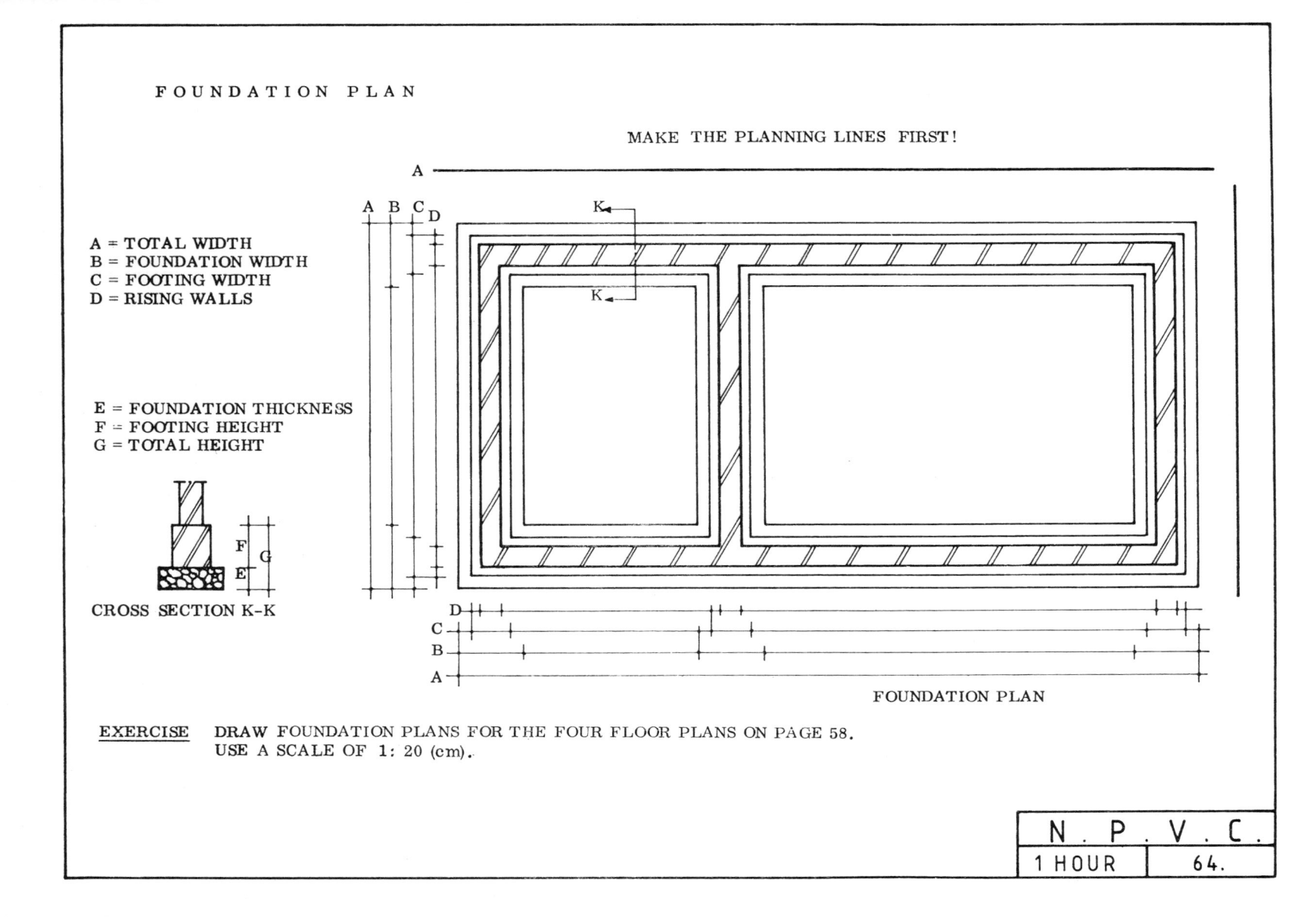

FOUNDATION PLAN

A = TOTAL WIDTH
B = FOUNDATION WIDTH
C = FOOTING WIDTH
D = RISING WALLS

E = FOUNDATION THICKNESS
F = FOOTING HEIGHT
G = TOTAL HEIGHT

CROSS SECTION K-K

EXERCISE DRAW FOUNDATION PLANS FOR THE FOUR FLOOR PLANS ON PAGE 58.
USE A SCALE OF 1: 20 (cm).

N.P.V.C.
1 HOUR | 64.

BUILDING WITH VERANDAH

PLANS - ELEVATIONS - CROSS SECTION

TECHNICAL DATA

FOOTING FOR THE VERANDAH IS SMALLER.

FOUNDATION FOR THE VERANDAH IS SMALLER.

VERANDAH FLOOR IS AT THE SAME LEVEL AS THE INSIDE FLOOR (X).

VERANDAH FLOOR IS PROJECTING ON THE PLAN (D).

NOTICE THE DIFFERENCE IN THE VERANDAH FLOOR AND THE FOOTINGS HEIGHT (K).

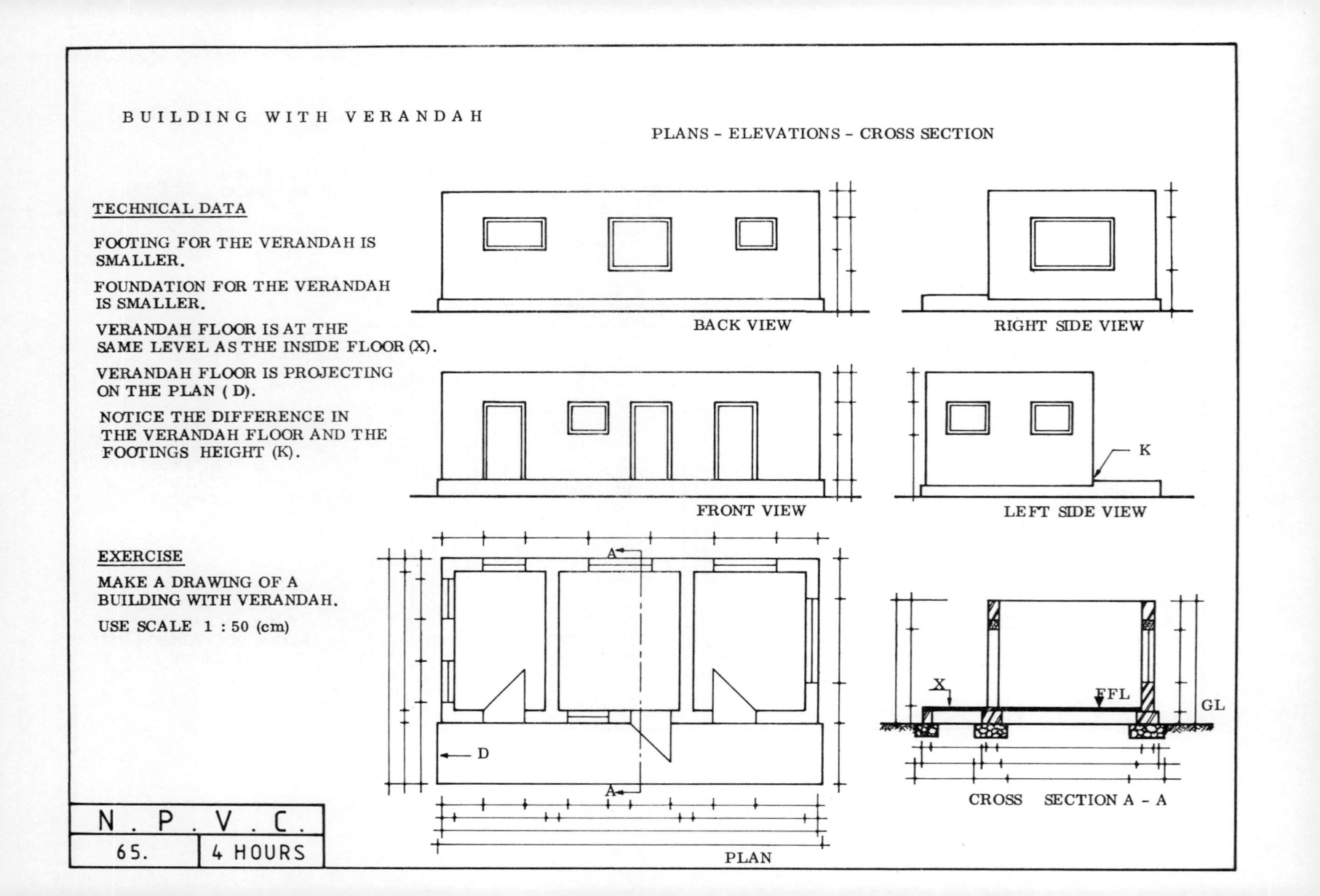

EXERCISE

MAKE A DRAWING OF A BUILDING WITH VERANDAH.

USE SCALE 1 : 50 (cm)

RIGHT SIDE VIEW

LEFT SIDE VIEW

CROSS SECTION A - A

BACK VIEW

FRONT VIEW

PLAN

N.P.V.C. 65. 4 HOURS

FOUNDATION PLAN FOR BUILDING WITH VERANDAH

NOTICE
K - K & N - N ARE CROSS SECTIONS.

DIMENSION LINES:
A = TOTAL LENGTH
B = FOUNDATIONS
C = FOOTINGS
D = WALLS

EXERCISE
MAKE FOUNDATION PLANS FOR THE PLANS ON PAGE 65.

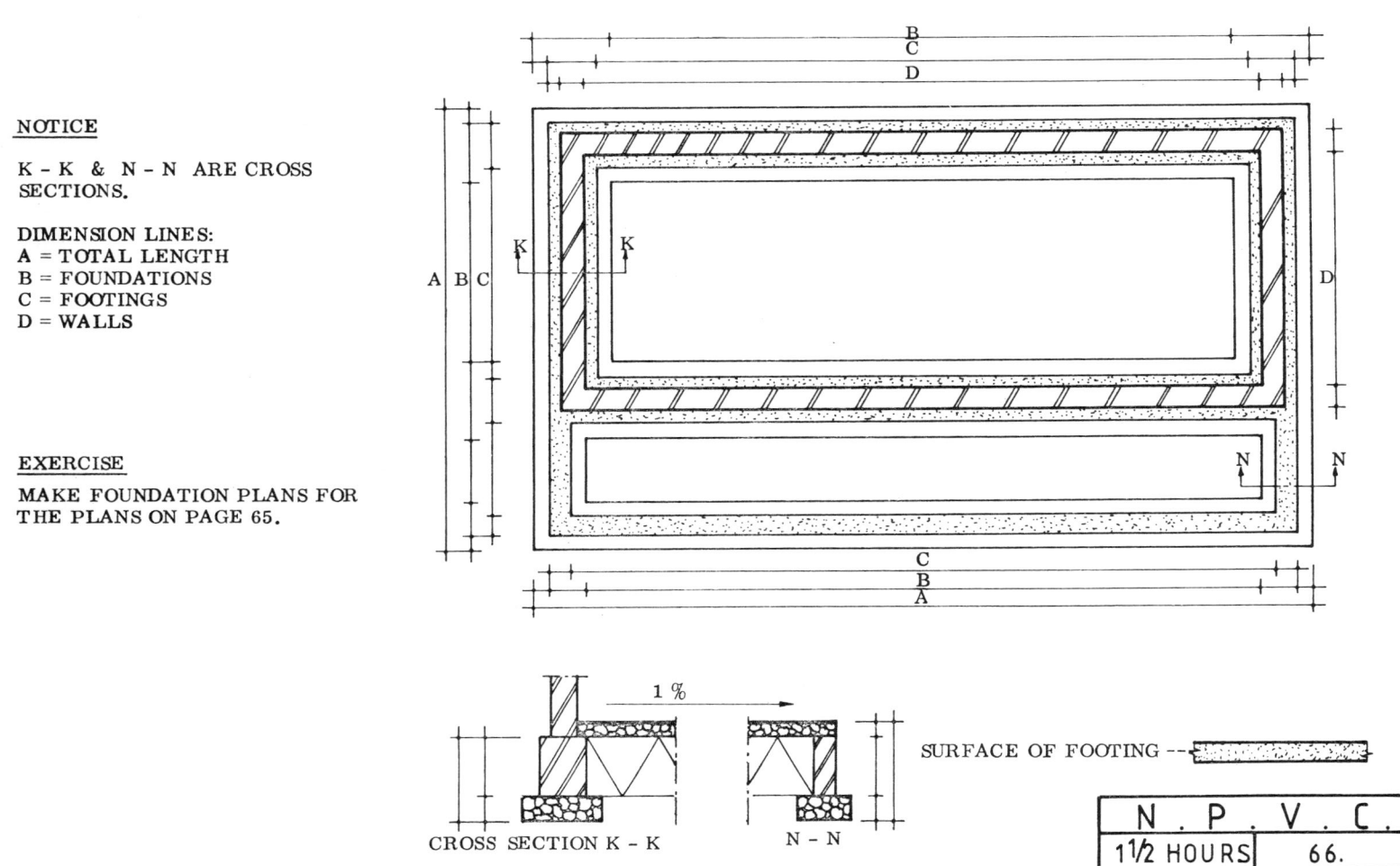

CROSS SECTION K - K

N - N

SURFACE OF FOOTING

1 %

N. P. V. C.
1½ HOURS | 66.

FRAME WITH REBATE

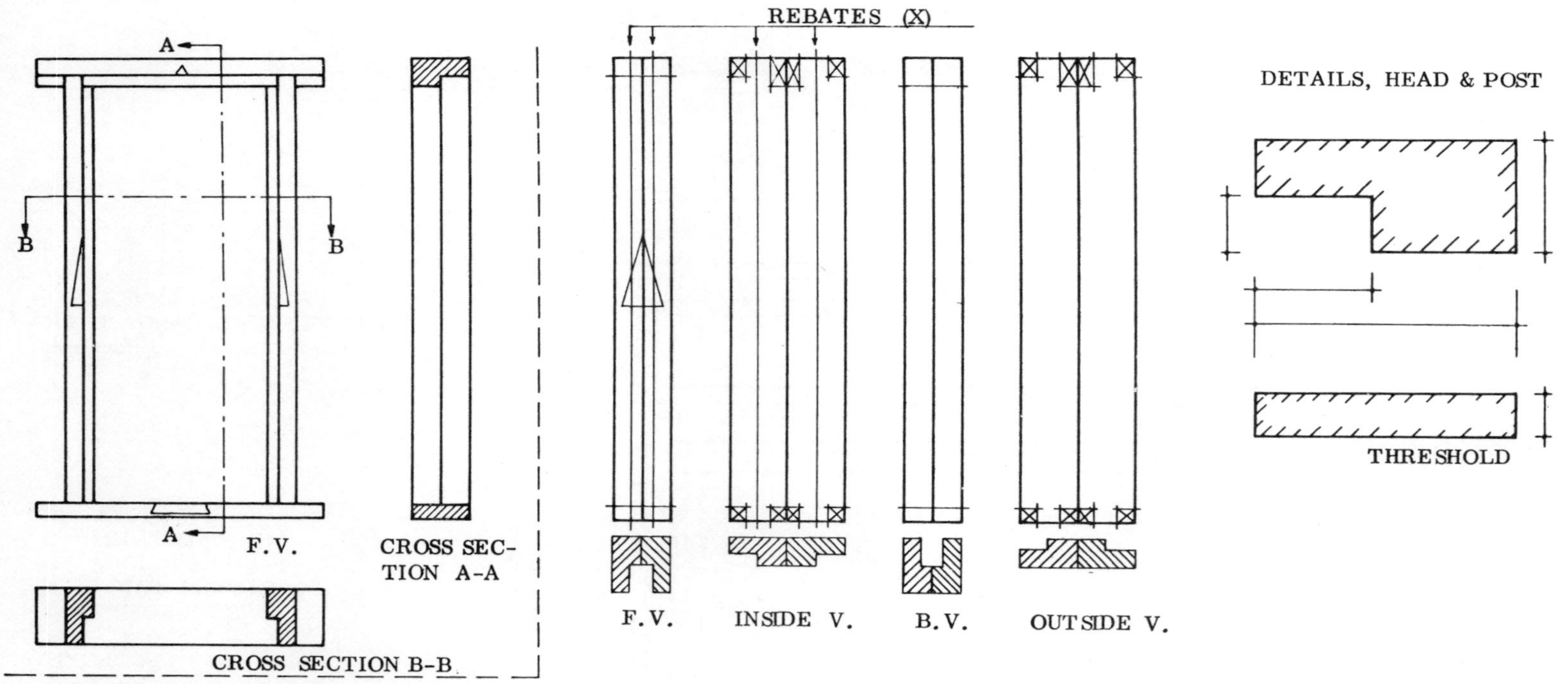

DETAILS, HEAD & POST

THRESHOLD

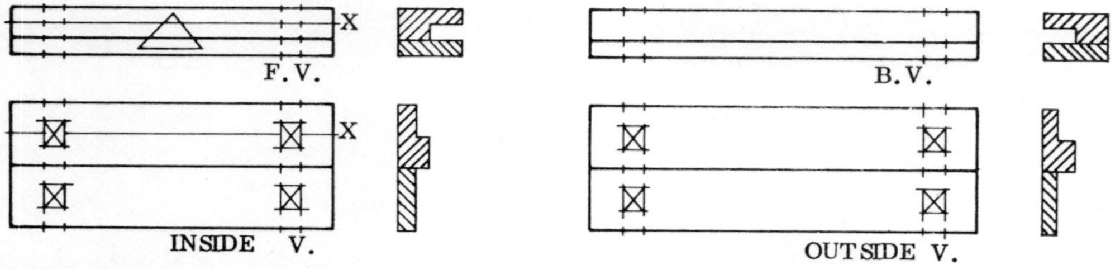

REBATES (X)

F.V.
INSIDE V.
B.V.
OUTSIDE V.

CROSS SECTION A-A

CROSS SECTION B-B

F.V.
INSIDE V.

B.V.
OUTSIDE V.

NOTICE
- CROSS SECTION A-A and B-B
  THERE IS NO T.V. or SIDE V.
- MARKING LINES FOR THE REBATES(X)
- STUDY THE CROSS SECTIONS OF THE
  MEMBERS(FOR MARKING OUT)
- SHOULDERS ARE DIFFERENT

EXERCISE
DRAW A DOORFRAME WITH THRESHOLD.
USE SCALE 1 : 10 (cm).
DRAW ALL MEMBERS AND MARK THEM
OUT FOR JOINTS.

N. P. V. C.

67.   2 ¾ HOURS

WHEN TO REDUCE MORTICE AND TENON (use teaching aids)

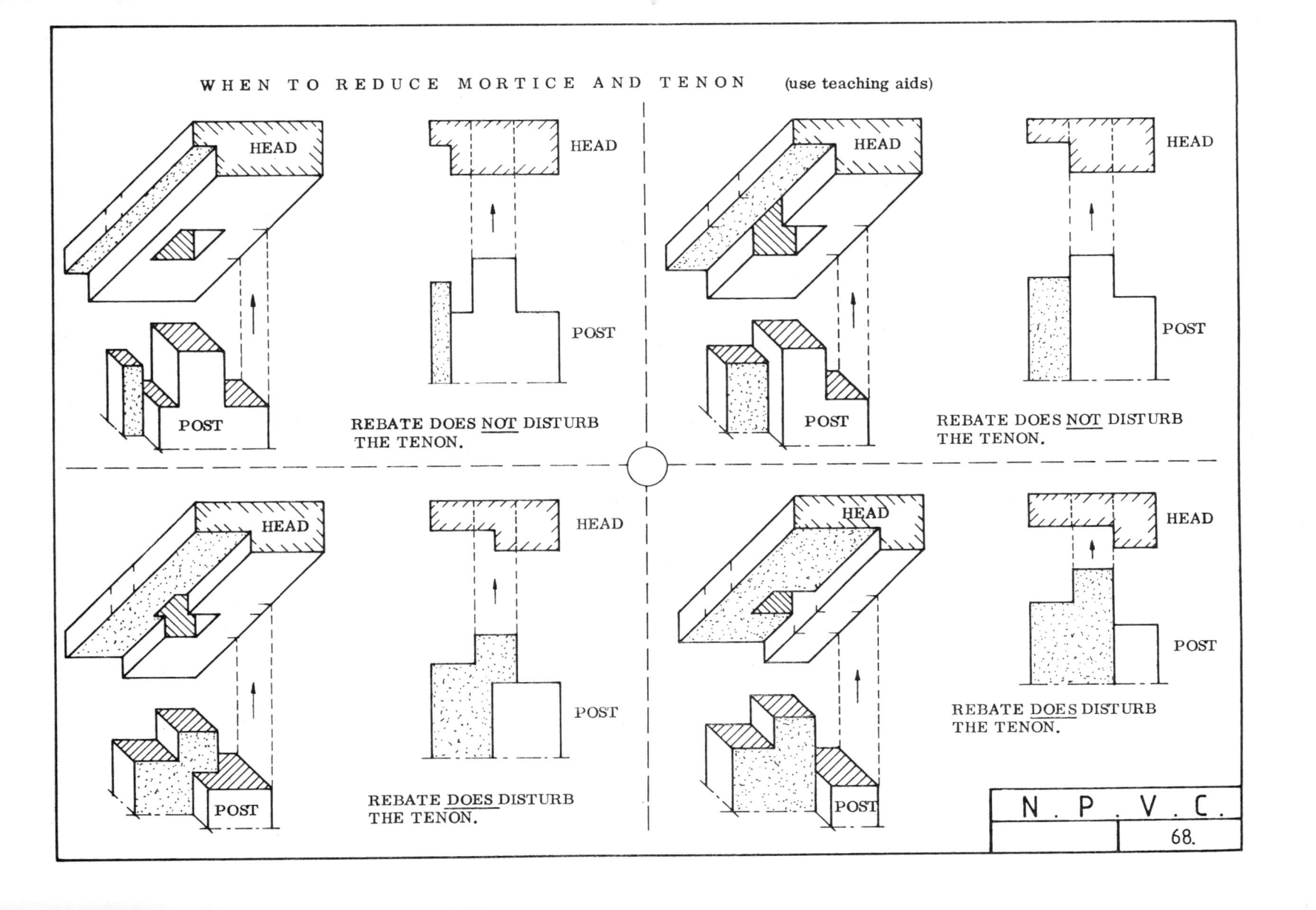

HEAD / POST

REBATE DOES NOT DISTURB THE TENON.

HEAD / POST

REBATE DOES DISTURB THE TENON.

HEAD / POST

REBATE DOES NOT DISTURB THE TENON.

HEAD / POST

REBATE DOES DISTURB THE TENON.

N.P.V.C.

68.

FRAME WITH CONCRETE SHOE

CR. S. A-A

F.V.

INSIDE V.

B.V.

OUTSIDE V.

CROSS SECTION B-B

DETAIL: DOWEL CONSTRUCTION

POST
SHOE
FLOOR

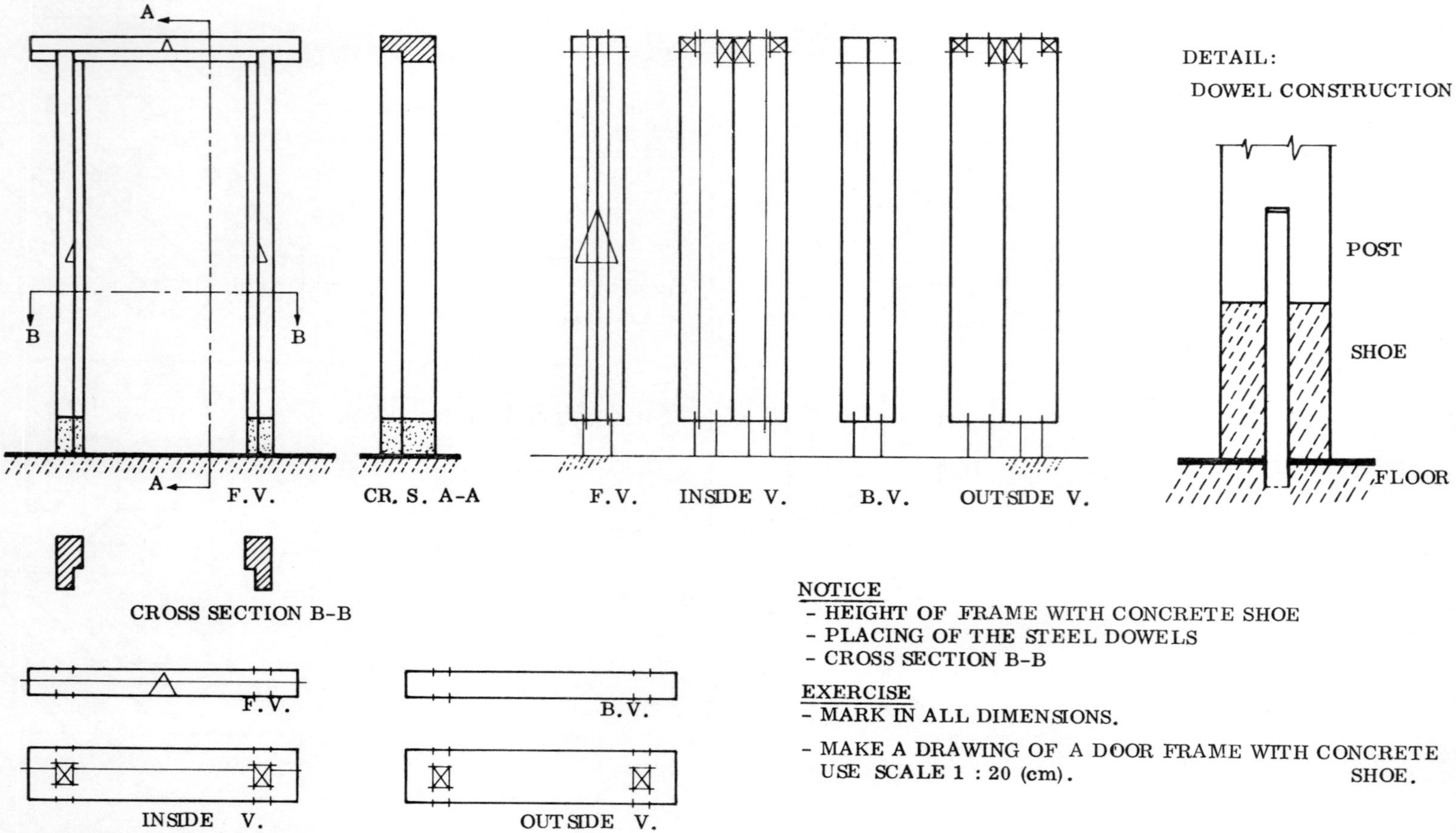

NOTICE
- HEIGHT OF FRAME WITH CONCRETE SHOE
- PLACING OF THE STEEL DOWELS
- CROSS SECTION B-B

EXERCISE
- MARK IN ALL DIMENSIONS.
- MAKE A DRAWING OF A DOOR FRAME WITH CONCRETE SHOE.
  USE SCALE 1 : 20 (cm).

F.V.
B.V.
INSIDE V.
OUTSIDE V.

N. P. V. C.
69.
2 ½ HOURS

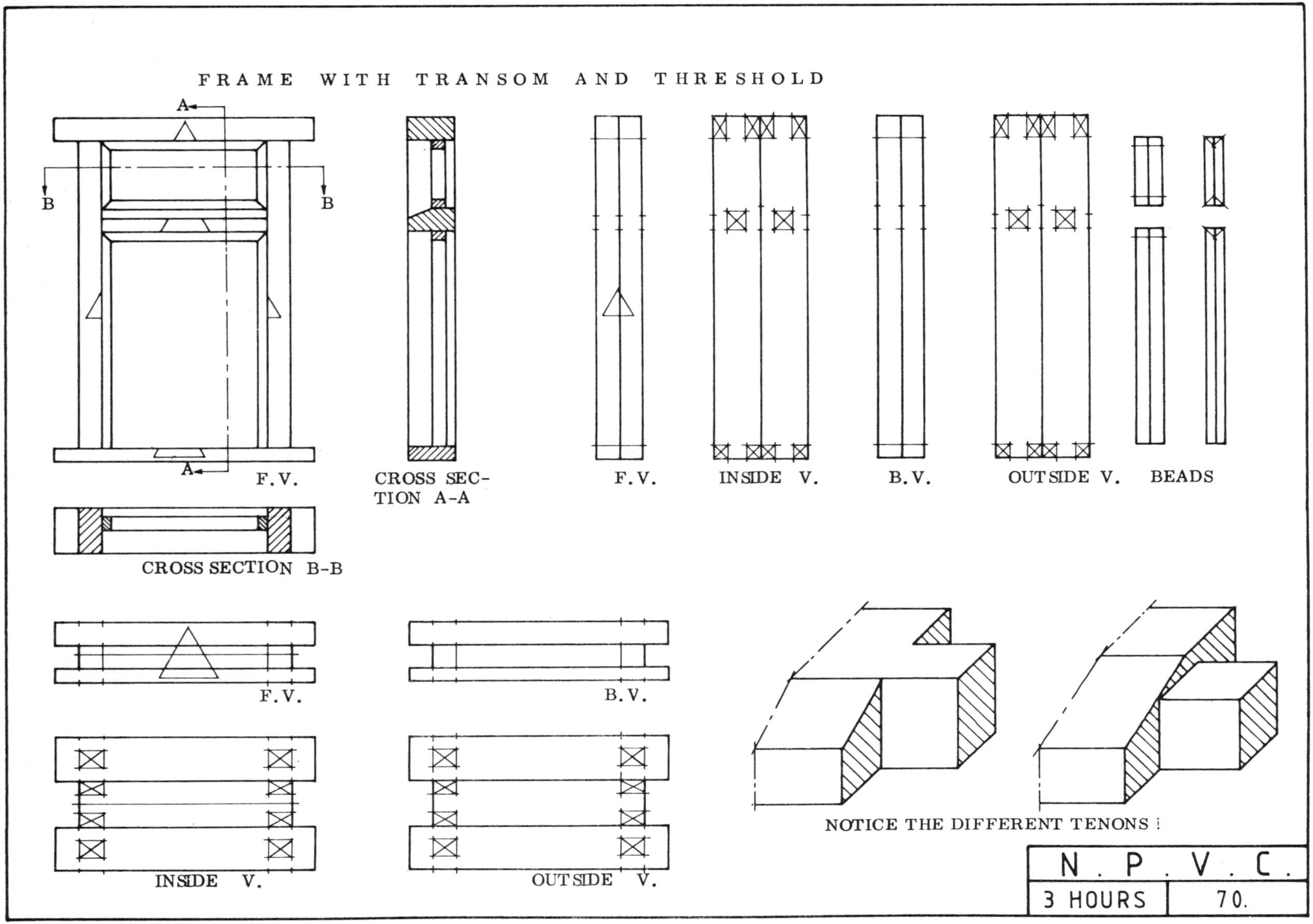

FRAME WITH TRANSOM AND THRESHOLD

OUTSIDE V. BEADS

B.V.

INSIDE V.

F.V.

CROSS SECTION A-A

CROSS SECTION B-B

F.V.

B.V.

OUTSIDE V.

F.V.

INSIDE V.

NOTICE THE DIFFERENT TENONS!

N. P. V. C.

3 HOURS | 70.

WHEN TO REDUCE MORTICE AND TENON

1
REBATE DOES NOT DISTURB THE TENON.

2
REBATE DOES NOT DISTURB THE TENON.

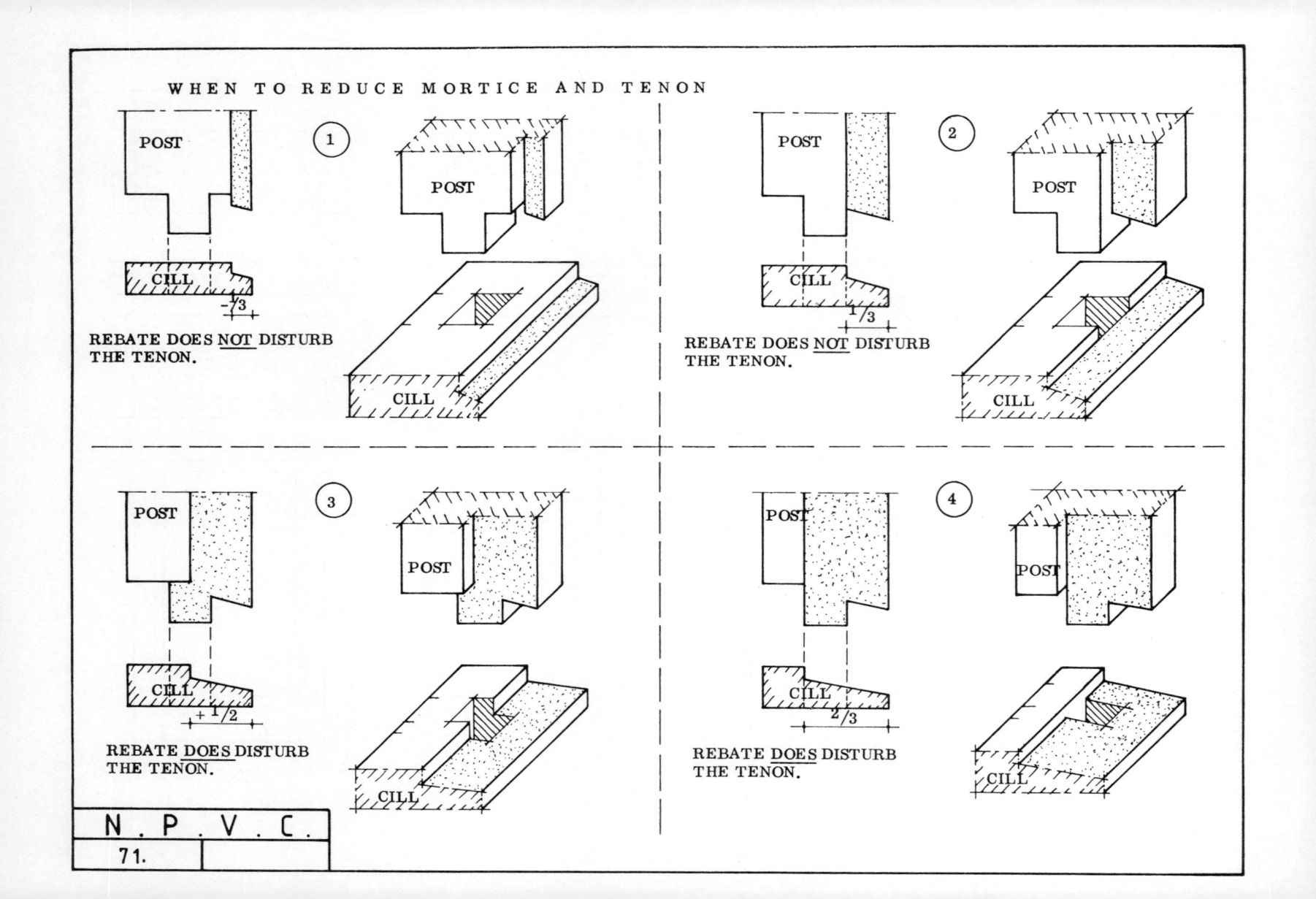

3
REBATE DOES DISTURB THE TENON.

4
REBATE DOES DISTURB THE TENON.

N.P.V.C.
71.

WINDOW FRAME WITH REBATE

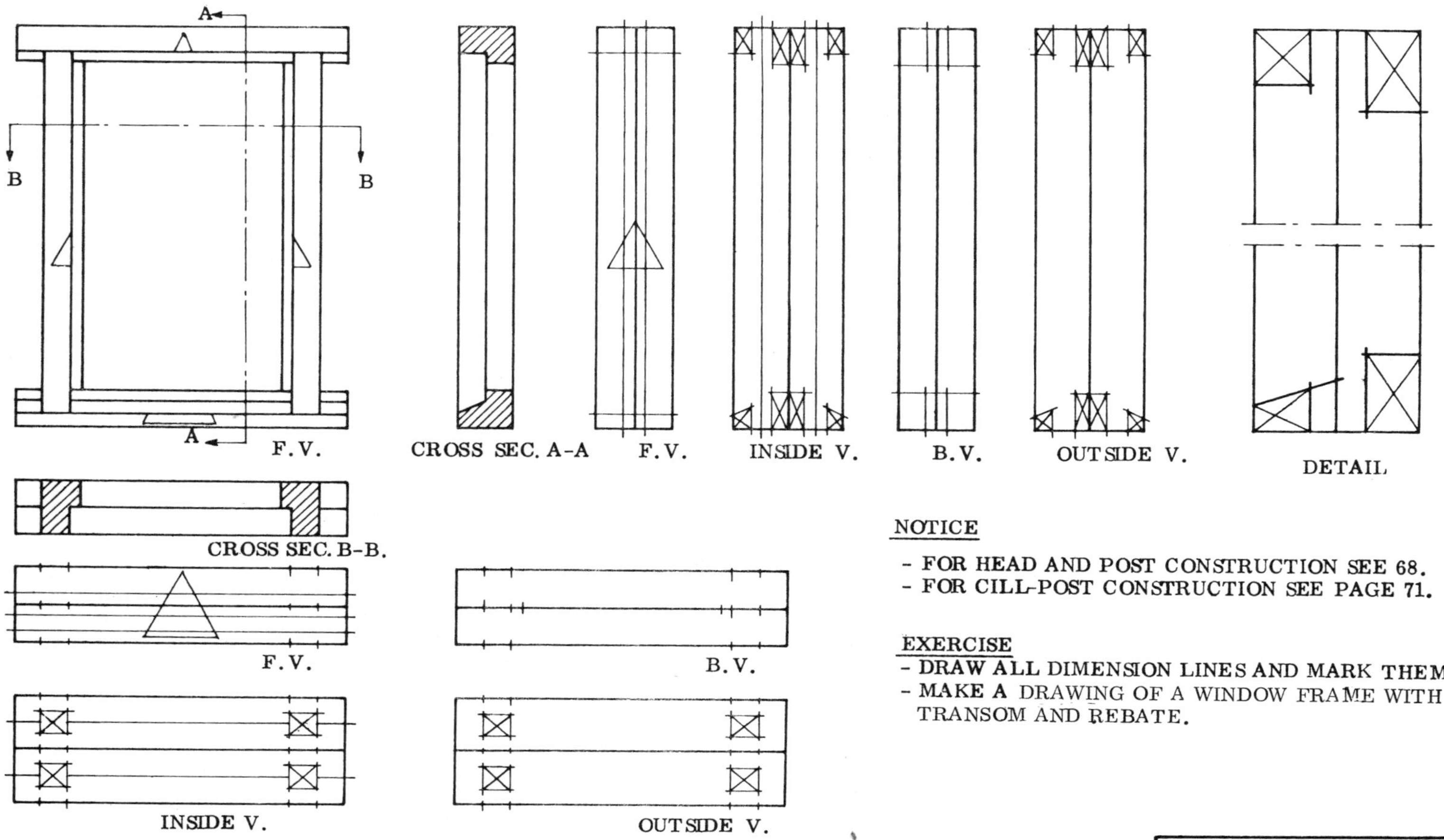

DETAIL.

OUTSIDE V.

B.V.

INSIDE V.

F.V.

CROSS SEC. A-A

CROSS SEC. B-B.

F.V.

B.V.

OUTSIDE V.

INSIDE V.

NOTICE
- FOR HEAD AND POST CONSTRUCTION SEE 68.
- FOR CILL-POST CONSTRUCTION SEE PAGE 71.

EXERCISE
- DRAW ALL DIMENSION LINES AND MARK THEM.
- MAKE A DRAWING OF A WINDOW FRAME WITH TRANSOM AND REBATE.

N. P. V. C.

2 ½ HOURS | 72.

LEDGED, BRACED AND BATTENED CASEMENT

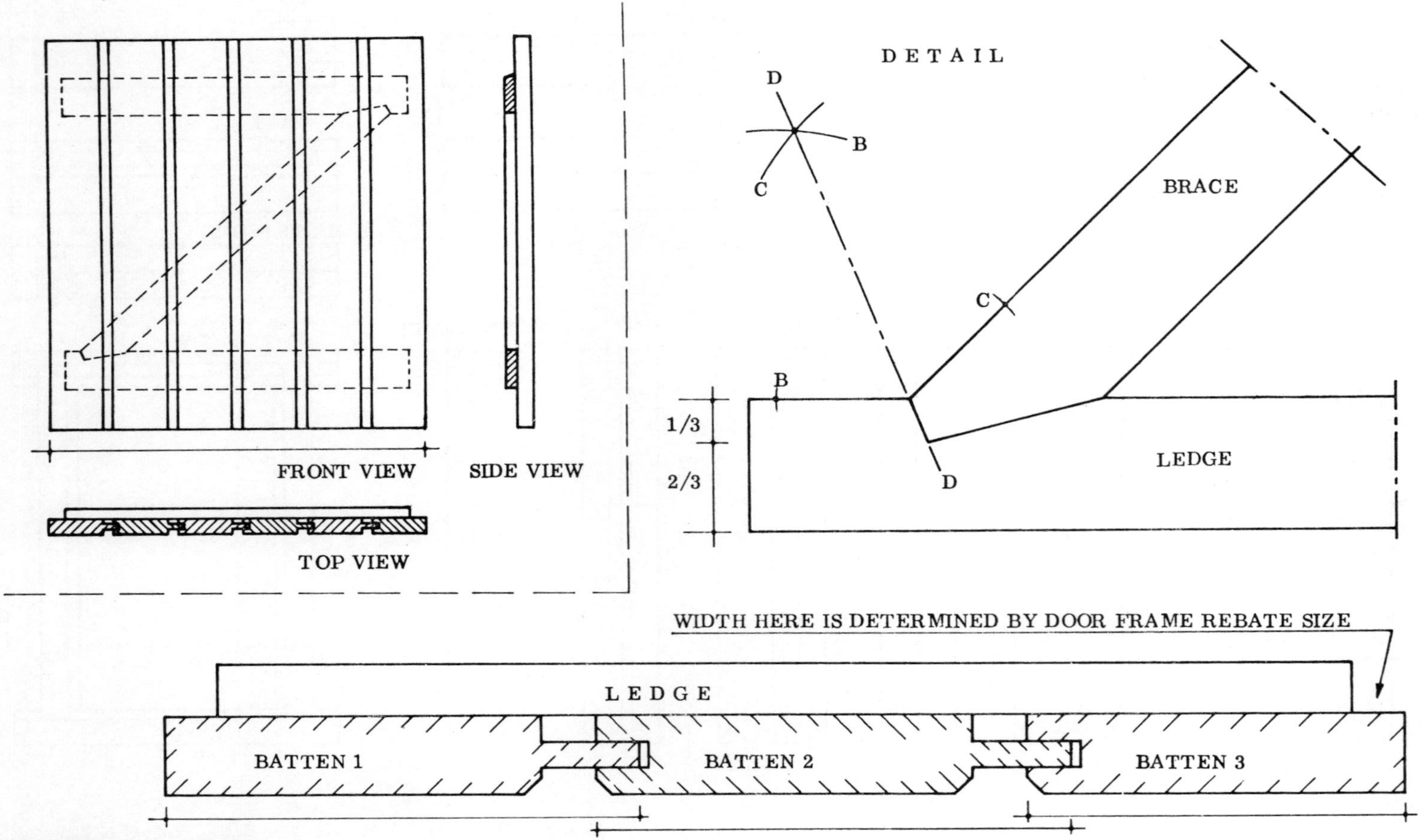

DETAIL

BRACE

LEDGE

1/3
2/3

FRONT VIEW

SIDE VIEW

TOP VIEW

WIDTH HERE IS DETERMINED BY DOOR FRAME REBATE SIZE

LEDGE

BATTEN 1

BATTEN 2

BATTEN 3

73.	N. P. V. C.
	2 HOURS

DETAIL

TOP RAIL

INTERMEDIATE RAIL

FLUSH CASEMENT

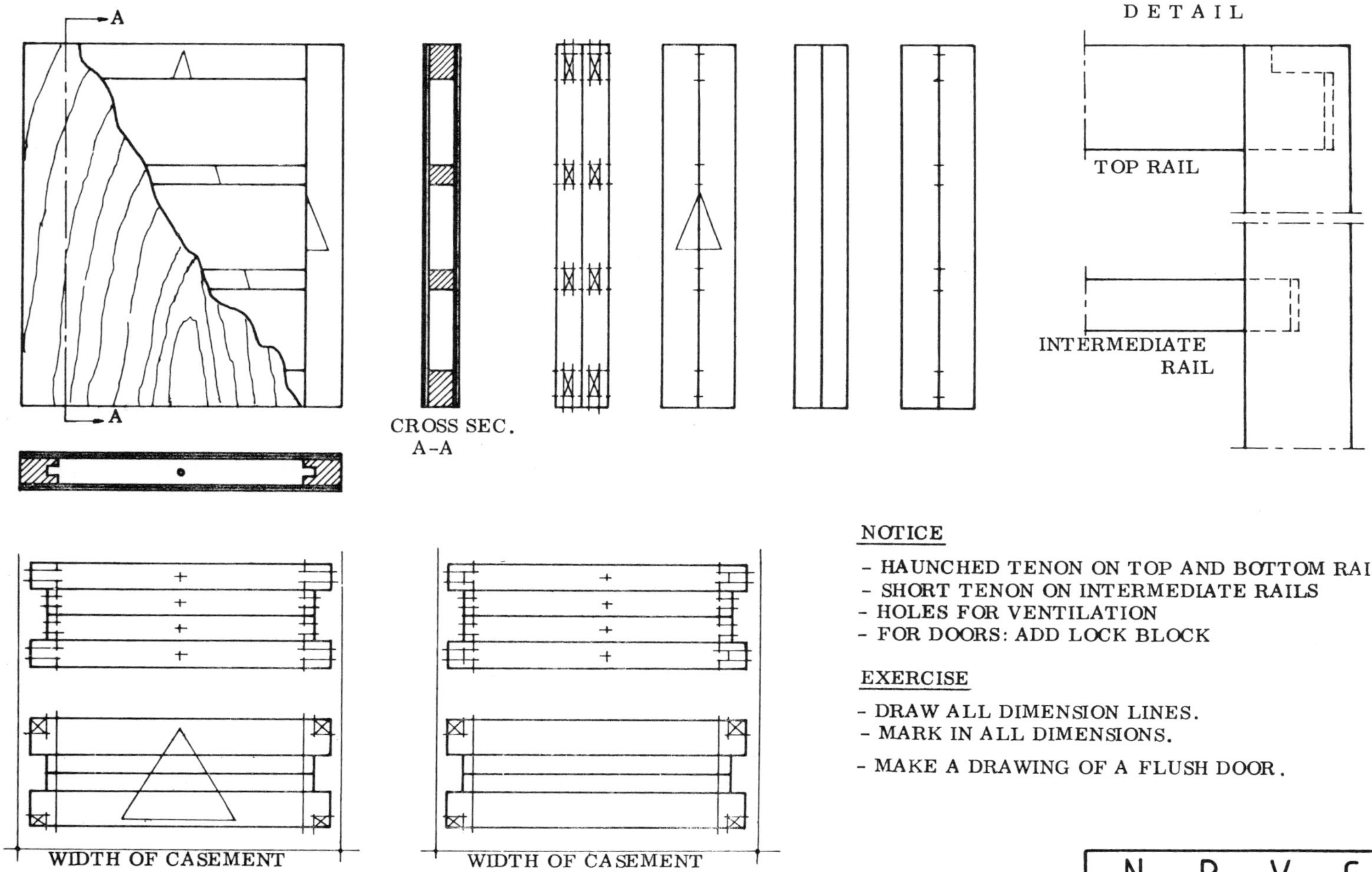

CROSS SEC. A-A

WIDTH OF CASEMENT

WIDTH OF CASEMENT

NOTICE
- HAUNCHED TENON ON TOP AND BOTTOM RAILS
- SHORT TENON ON INTERMEDIATE RAILS
- HOLES FOR VENTILATION
- FOR DOORS: ADD LOCK BLOCK

EXERCISE
- DRAW ALL DIMENSION LINES.
- MARK IN ALL DIMENSIONS.
- MAKE A DRAWING OF A FLUSH DOOR.

N.P.V.C.

2 ¾ HOURS

74.

PANELLED CASEMENT

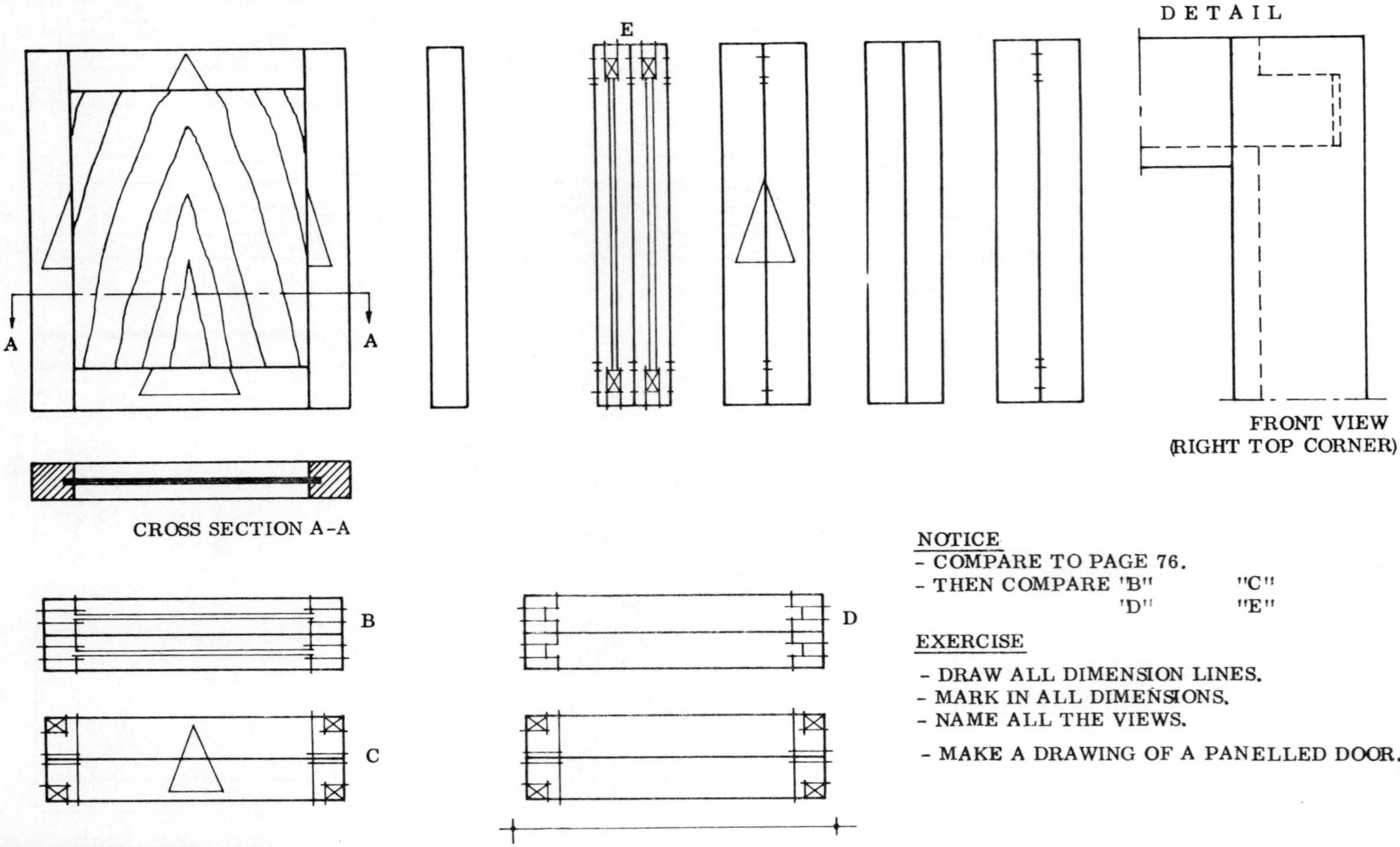

DETAIL

FRONT VIEW
(RIGHT TOP CORNER)

CROSS SECTION A-A

B
C
D
E

NOTICE:
- COMPARE TO PAGE 76.
- THEN COMPARE "B" "C"
   "D" "E"

EXERCISE:
- DRAW ALL DIMENSION LINES.
- MARK IN ALL DIMENSIONS.
- NAME ALL THE VIEWS.
- MAKE A DRAWING OF A PANELLED DOOR.

REMEMBER THAT THE TOP AND BOTTOM RAILS ARE SHORTER BECAUSE OF THE STOPPED MORTICE AND TENON !

N. P. V. C.
75. 2 ½ HOURS

GROOVED STILE AND TOPRAIL MARKING OUT

HAUNCHING
MORTICE
REDUCED WIDTH

INSIDE VIEW "E"

STILE

OBLIQUE VIEW

NOTICE DOTTED AREAS

OUTSIDE VIEW "D"

TOP RAIL

SIDE VIEW "C"

INSIDE VIEW "B"

N.P.V.C.

76.

REBATED CASEMENT

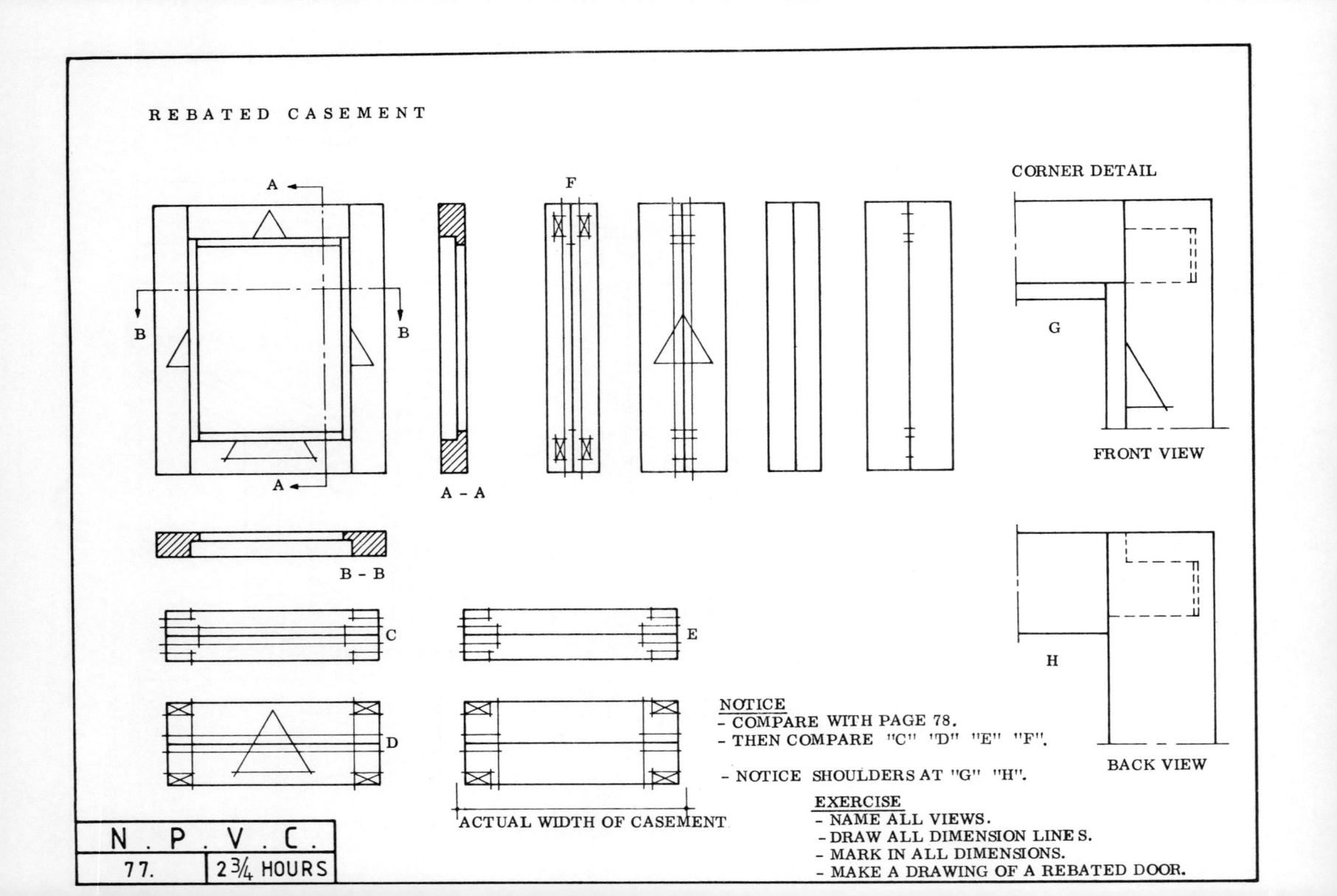

CORNER DETAIL

FRONT VIEW

BACK VIEW

G

H

A – A

B – B

C

D

E

F

ACTUAL WIDTH OF CASEMENT

NOTICE
- COMPARE WITH PAGE 78.
- THEN COMPARE "C" "D" "E" "F".
- NOTICE SHOULDERS AT "G" "H".

EXERCISE
- NAME ALL VIEWS.
- DRAW ALL DIMENSION LINES.
- MARK IN ALL DIMENSIONS.
- MAKE A DRAWING OF A REBATED DOOR.

N. P. V. C.

77. 2¾ HOURS

REBATED STILE AND TOPRAIL CONSTRUCTION

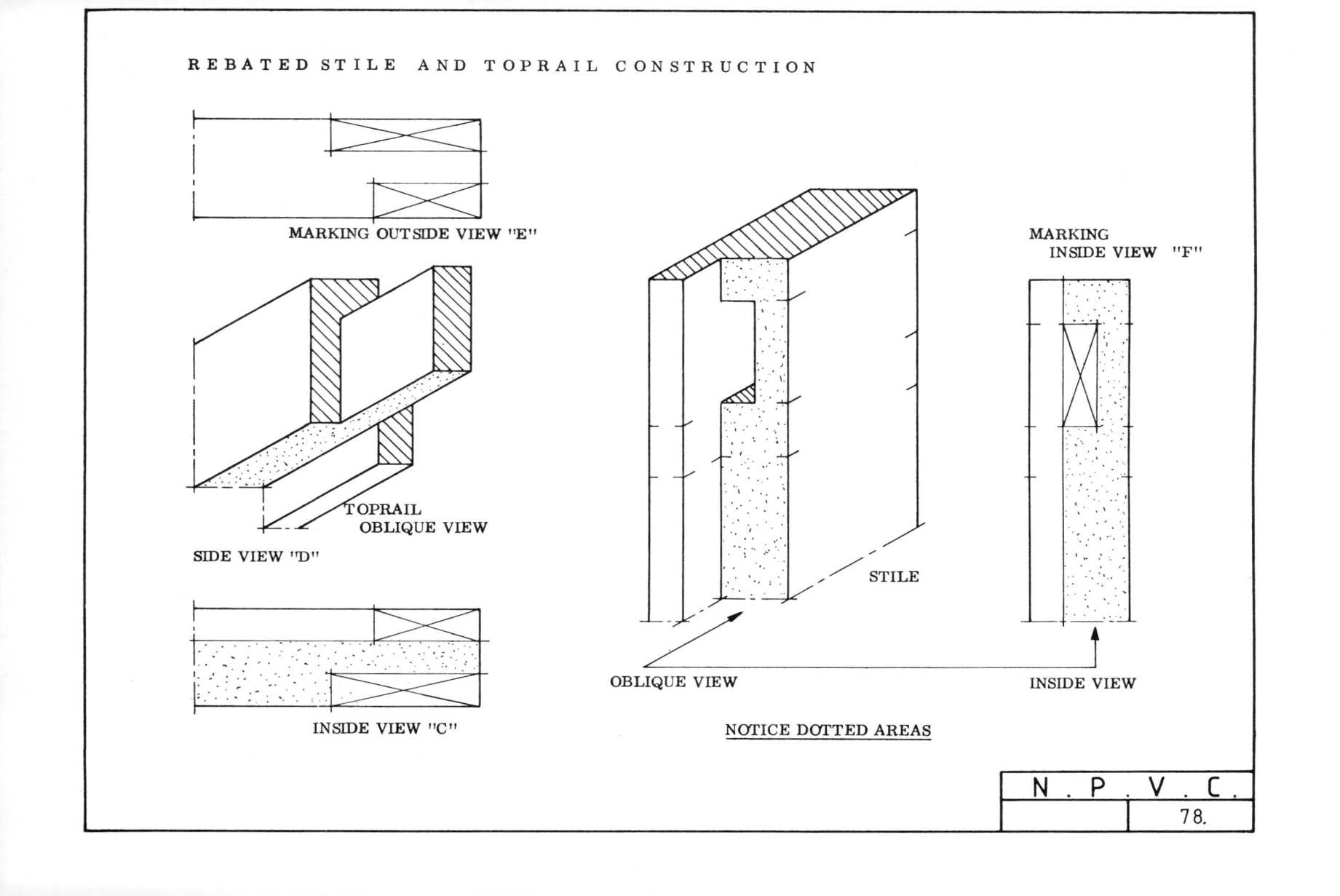

MARKING INSIDE VIEW "F"

INSIDE VIEW

STILE

OBLIQUE VIEW

NOTICE DOTTED AREAS

MARKING OUTSIDE VIEW "E"

TOPRAIL OBLIQUE VIEW

SIDE VIEW "D"

INSIDE VIEW "C"

N.P.V.C.

78.

WORKING DRAWINGS

TO BE ABLE TO MAKE A WORKING DRAWING FOR A BUILDING YOU NEED A SKETCHPLAN WITH THE NECESSARY DETAILS SUCH AS INSIDE DIMENSIONS OF THE ROOMS, THE POSITIONS OF THE WINDOWS AND DOORS, ETC. ON THE NEXT PAGE YOU WILL FIND A LAY-OUT OF HOW TO WRITE DOWN ALL THE TECHNICAL DATA REGARDING MATERIALS AND SOME DIMENSIONS.

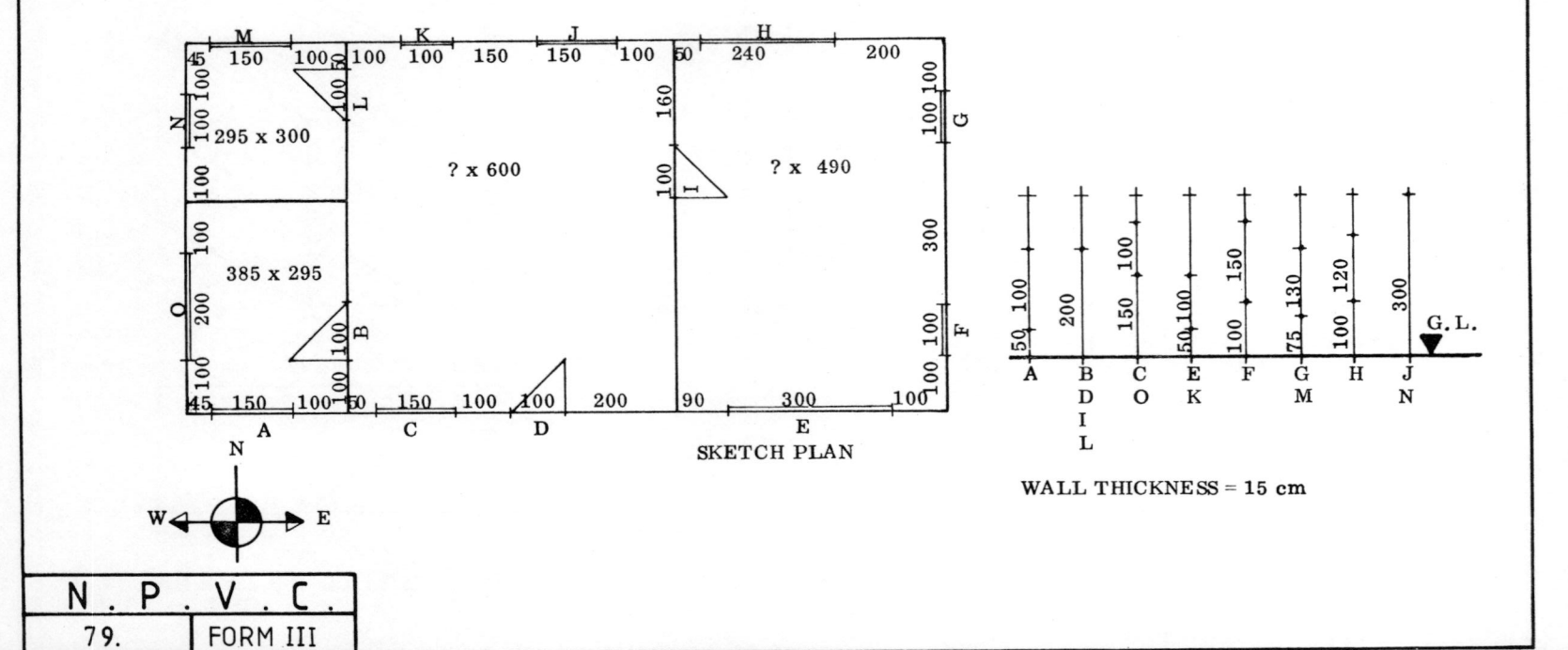

SKETCH PLAN

WALL THICKNESS = 15 cm

N.P.V.C. — FORM III — 79.

TECHNICAL DATA

ROOF CONSTRUCTION, MAIN BUILDING

KIND OF ROOF			ODUM WOOD
RISE OF ROOF		cm	
RISE OF TRUSS		cm	
KIND OF COVERING	x	cm	ALUMINIUM
SPAN OF TRUSS		cm	
ROOF OVERHANG, LONG SIDES		cm	
ROOF PROJECTION AT GABLE		cm	
CEILING THICKNESS		cm	PLYWOOD
FASCIA BOARDS	x	cm	WAWA
RAFTERS	x	cm	ODUM WOOD
PURLINS	x	cm	,, ,,
TIE BEAM	x	cm	,, ,,
BRACES	x	cm	,, ,,
WALL PLATE	x	cm	,, ,,

VERANDAH, ROOF CONSTRUCTION

KIND OF ROOF			ODUM WOOD
RISE OF ROOF		cm	ODUM WOOD
SPAN OF TRUSS		cm	
OVERHANG ON LONG SIDES		cm	
PROJECTION AT GABLE ENDS		cm	PLYWOOD
CEILING THICKNESS		cm	WAWA
FASCIA BOARDS	x	cm	ODUM WOOD
RAFTERS	x	cm	
PURLINS	x	cm	,, ,,
TIE BEAM	x	cm	,, ,,
BRACES	x	cm	,, ,,

MAIN BUILDING

FOUNDATION	x	cm	CONCRETE
FOOTING HEIGHT ABOVE G.L.		cm	SANDCRETE
FOOTING WIDTH		cm	
WALL THICKNESS		cm	LANDCRETE
PLASTER THICKNESS (inside)		cm	CEMENT PLASTER
PLASTER THICKNESS (outside)		cm	,, ,,
FLOOR THICKNESS		cm	ONE COURSE WORK

VERANDAH MAIN BUILDING

COLUMNS	x	cm	REINFORCED CONCRETE
EAVE BEAM	x	cm	REINFORCED CONCRETE
EAVE PLATE	x	cm	ODUM WOOD
FLOOR SLOPE		%	PER METER
FOOTING HEIGHT ABOVE G.L.		cm	
FOOTING WIDTH		cm	SANDCRETE
FOUNDATION	x	cm	CONCRETE
FLOOR THICKNESS		cm	ONE COURSE WORK

DIRECTION OF BUILDING   EAST / WEST   NORTH – SOUTH

SPECIAL REQUIREMENTS

N.P.V.C.

80.

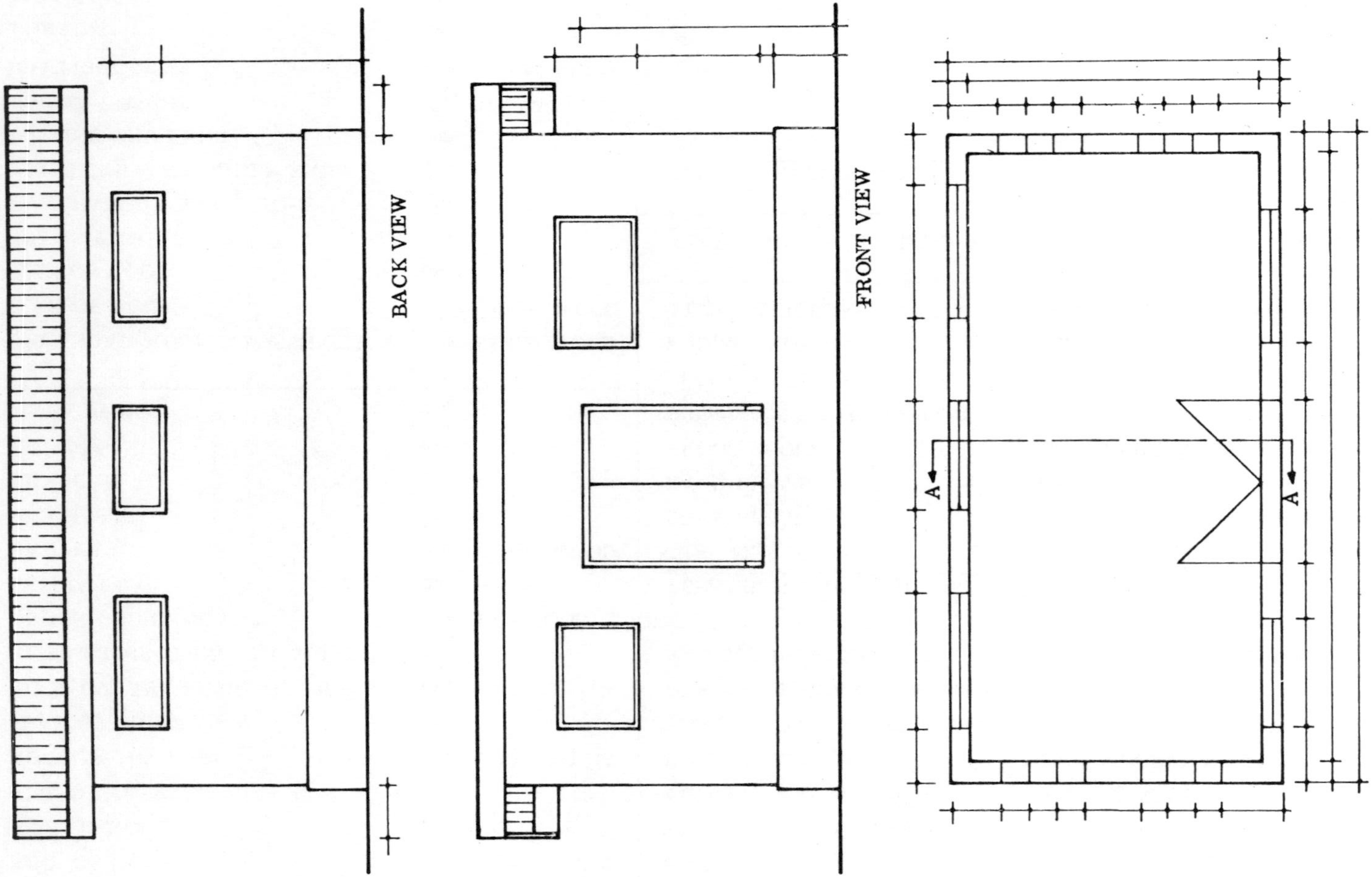

PLAN

FRONT VIEW

BACK VIEW

EXERCISE: MAKE AN ORTHOGRAPHIC DRAWING WITH CROSS SECTION OF A BUILDING WITH PENTROOF. USE SCALE 1 : 100 (cm).

| 81. | N. P. V. C. | 5 HOURS |

BUILDING WITH PENTROOF

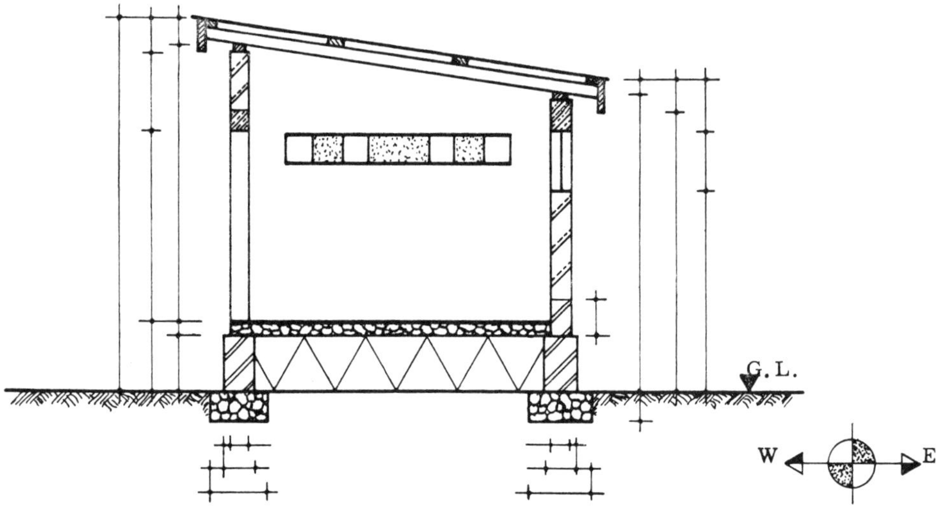

N.P.V.C.
82.

CROSS SECTION A-A

LEFT SIDE VIEW

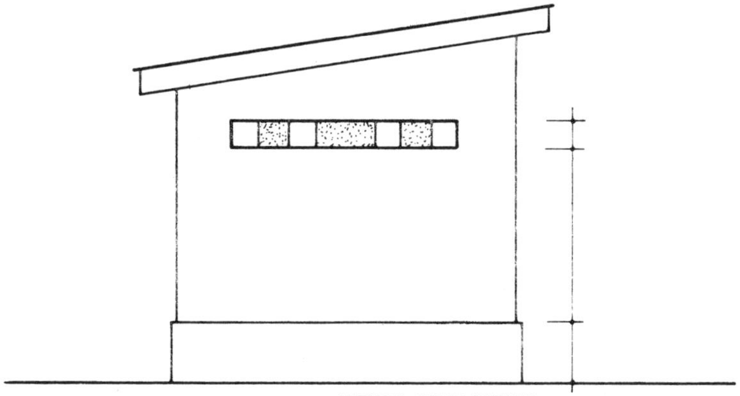

RIGHT SIDE VIEW

TEAR OUT THIS PAGE AND GLUE IT TO PAGE 81

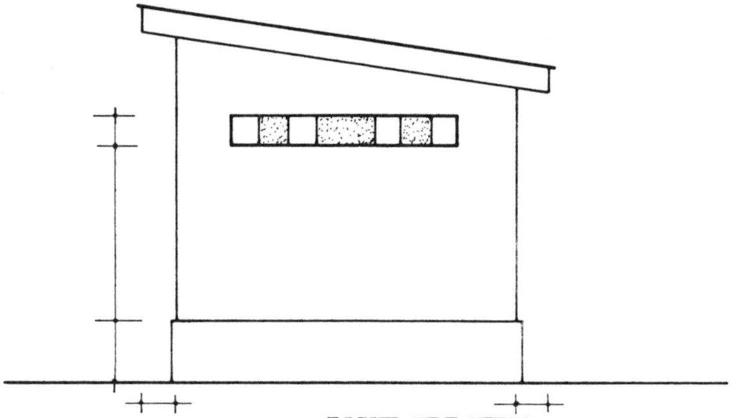

PENT ROOF PLAN

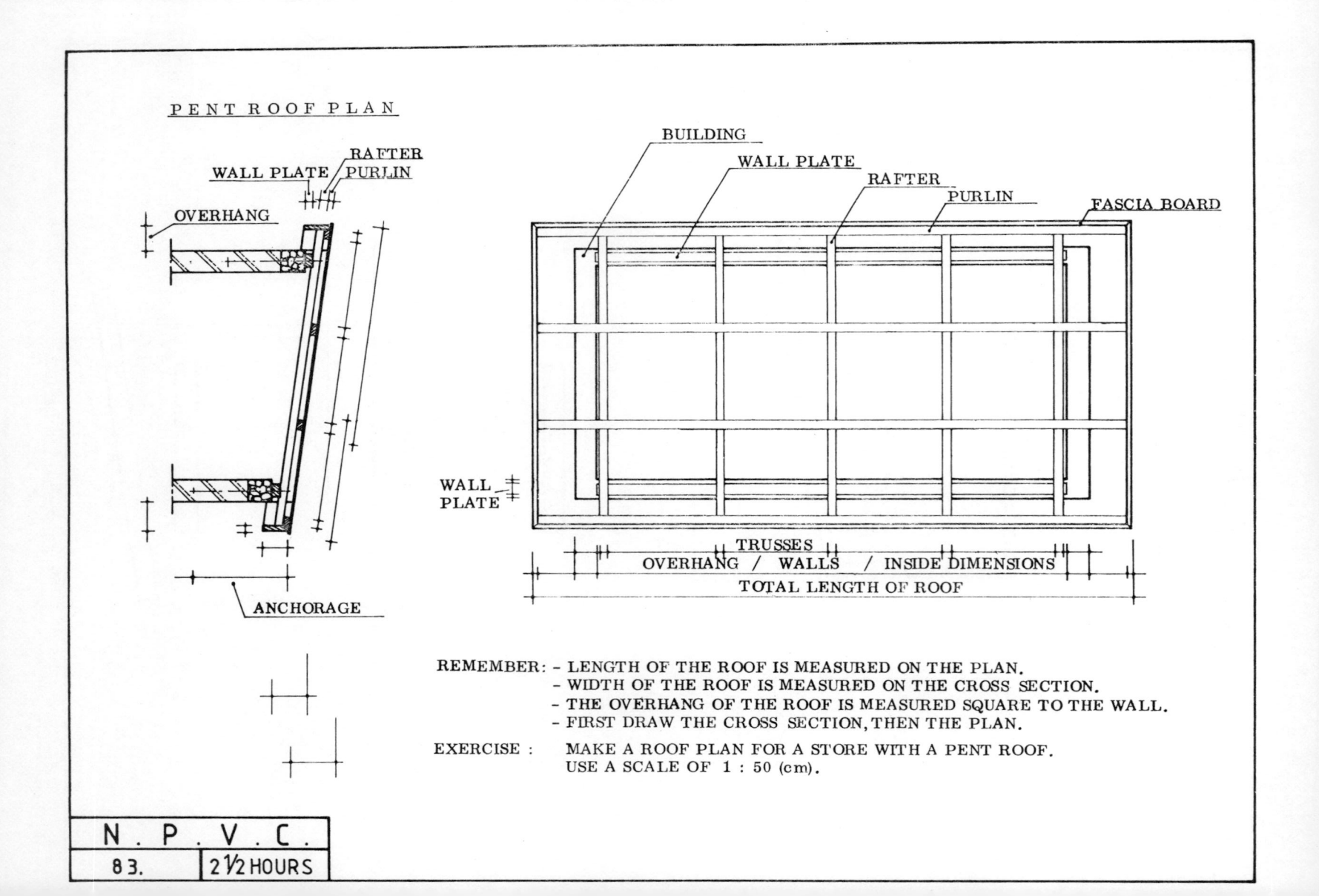

Labels: FASCIA BOARD, PURLIN, RAFTER, WALL PLATE, BUILDING, TRUSSES, WALL PLATE, OVERHANG / WALLS / INSIDE DIMENSIONS, TOTAL LENGTH OF ROOF

Labels (cross section): RAFTER, PURLIN, WALL PLATE, OVERHANG, ANCHORAGE

REMEMBER:
- LENGTH OF THE ROOF IS MEASURED ON THE PLAN.
- WIDTH OF THE ROOF IS MEASURED ON THE CROSS SECTION.
- THE OVERHANG OF THE ROOF IS MEASURED SQUARE TO THE WALL.
- FIRST DRAW THE CROSS SECTION, THEN THE PLAN.

EXERCISE : MAKE A ROOF PLAN FOR A STORE WITH A PENT ROOF. USE A SCALE OF 1 : 50 (cm).

N. P. V. C.

83. | 2½ HOURS

PARAPETTED PENT ROOF

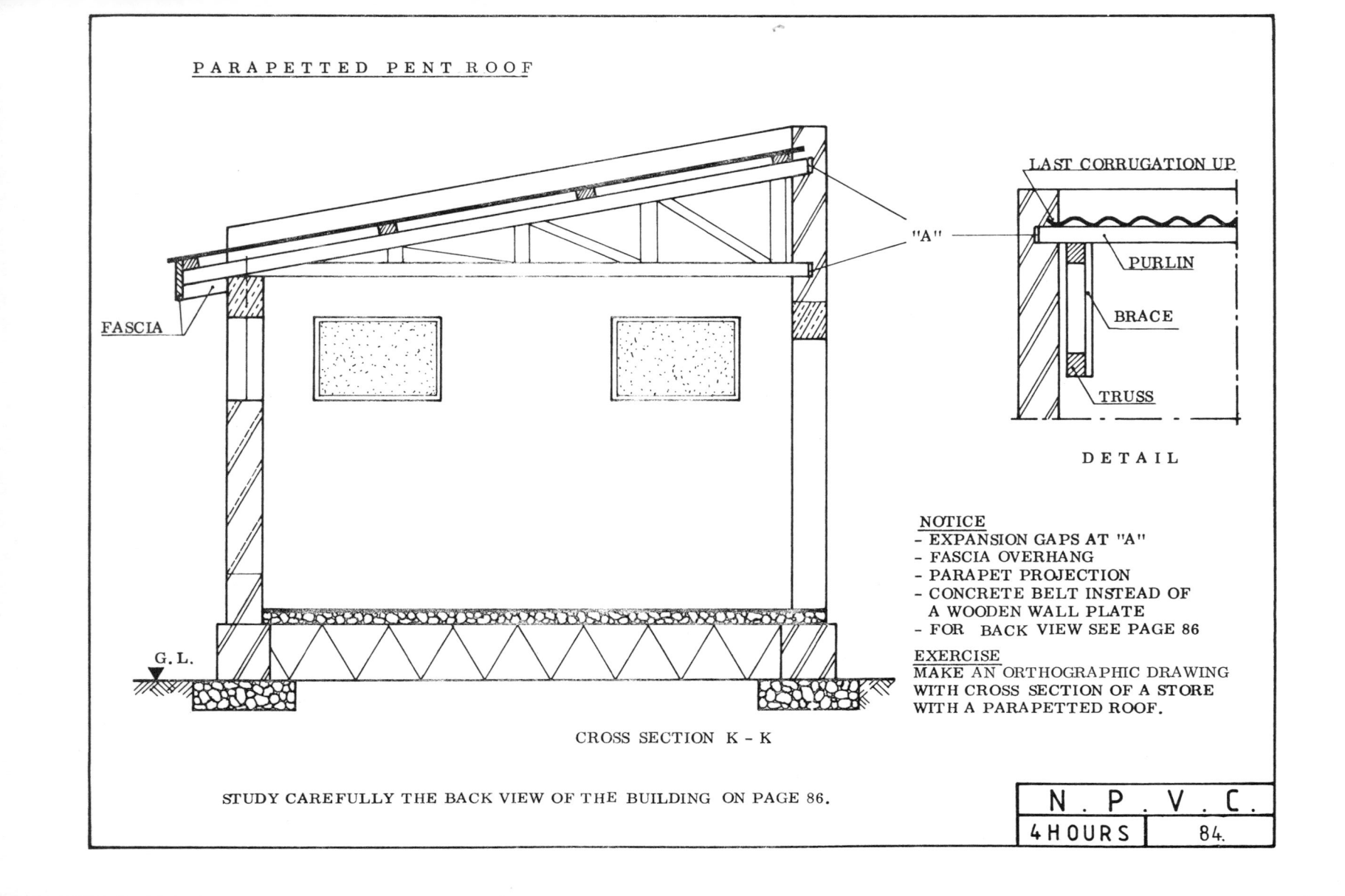

LAST CORRUGATION UP
PURLIN
BRACE
TRUSS

DETAIL

"A"

FASCIA

G.L.

CROSS SECTION K - K

NOTICE
- EXPANSION GAPS AT "A"
- FASCIA OVERHANG
- PARAPET PROJECTION
- CONCRETE BELT INSTEAD OF A WOODEN WALL PLATE
- FOR BACK VIEW SEE PAGE 86

EXERCISE
MAKE AN ORTHOGRAPHIC DRAWING WITH CROSS SECTION OF A STORE WITH A PARAPETTED ROOF.

STUDY CAREFULLY THE BACK VIEW OF THE BUILDING ON PAGE 86.

N.P.V.C.
4 HOURS
84.

PARAPETTED GABLE ROOF

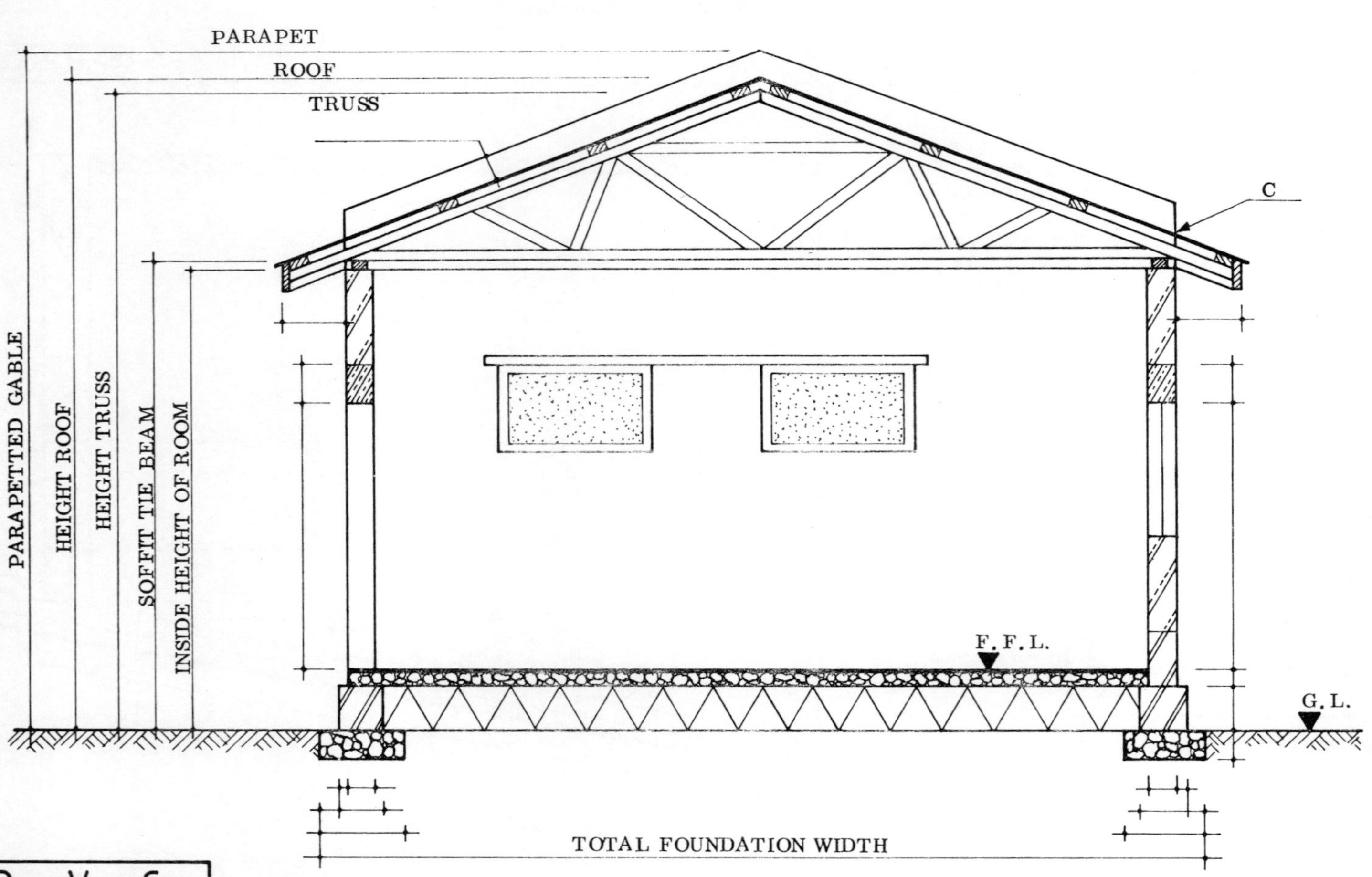

C

PARAPET
ROOF
TRUSS

G.L.

F.F.L.

TOTAL FOUNDATION WIDTH

INSIDE HEIGHT OF ROOM
SOFFIT TIE BEAM
HEIGHT TRUSS
HEIGHT ROOF
PARAPETTED GABLE

N. P. V. C.
85.

BACK VIEW OF A BUILDING WITH GABLE ROOF

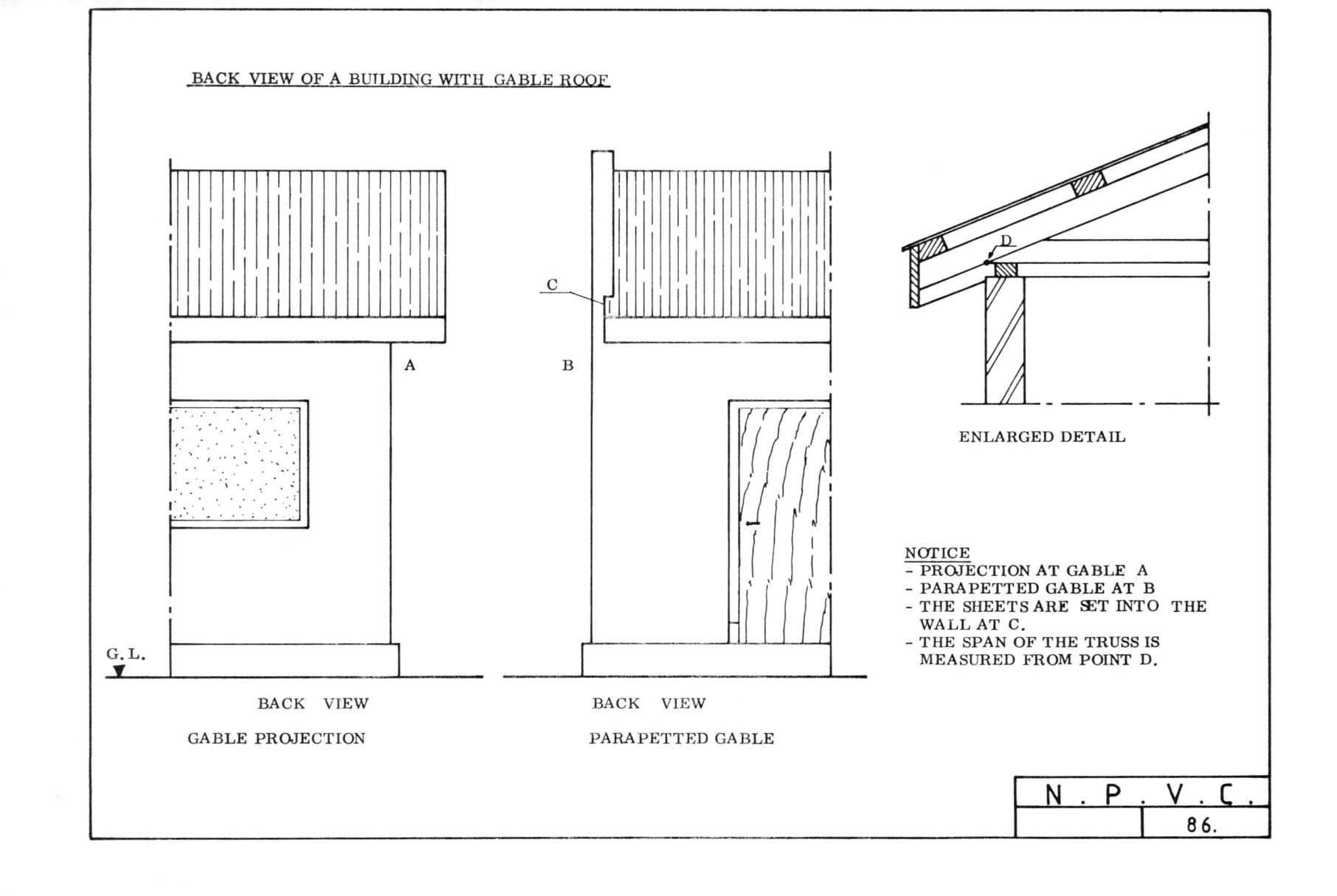

ENLARGED DETAIL

NOTICE
- PROJECTION AT GABLE A
- PARAPETTED GABLE AT B
- THE SHEETS ARE SET INTO THE WALL AT C.
- THE SPAN OF THE TRUSS IS MEASURED FROM POINT D.

BACK VIEW
PARAPETTED GABLE

BACK VIEW
GABLE PROJECTION

G.L.

N.P.V.C.
86.

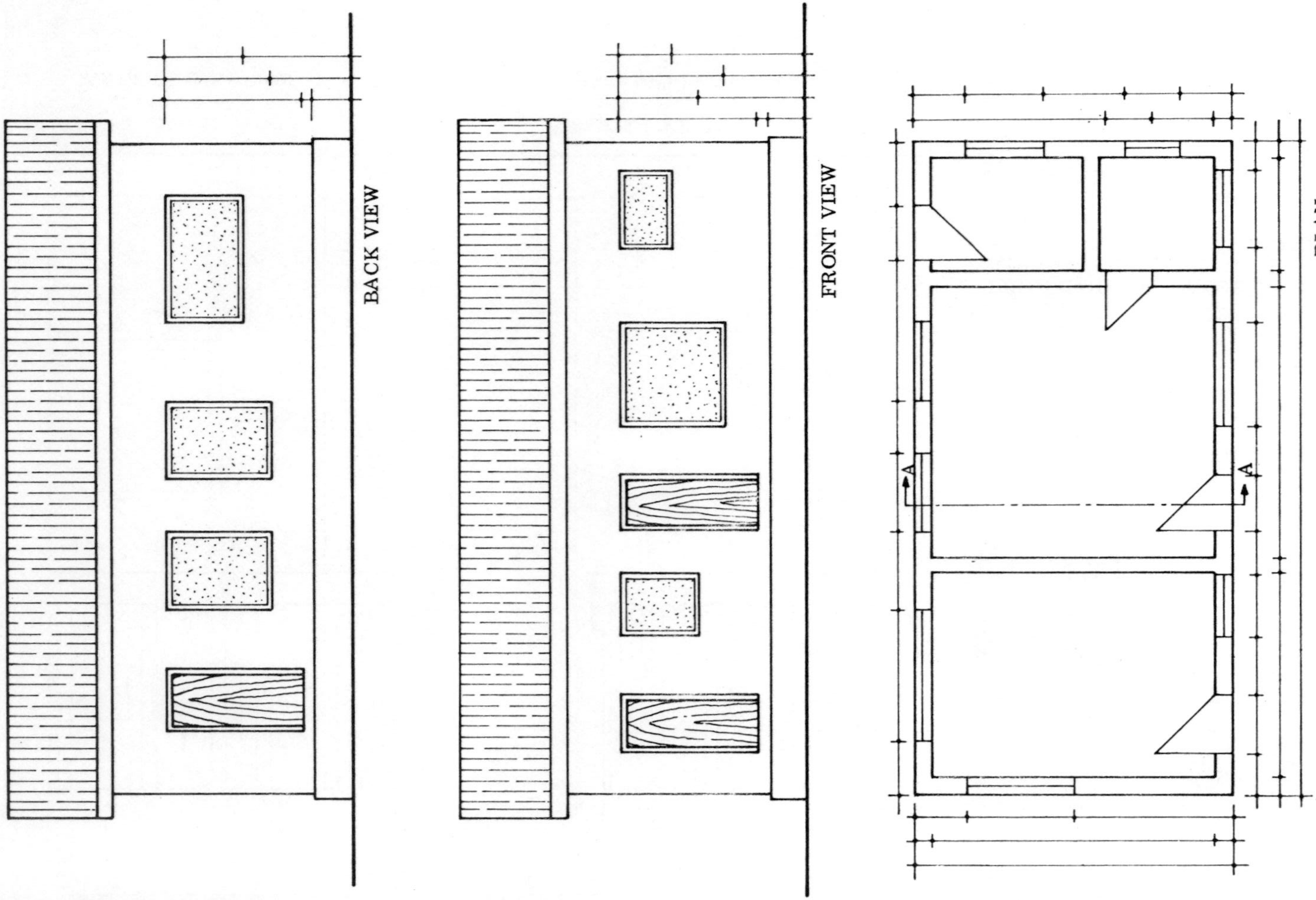

PLAN

FRONT VIEW

BACK VIEW

EXERCISE: MAKE AN ORTHOGRAPHIC DRAWING WITH CROSS SECTION OF A BUILDING WITH A GABLE ROOF; USE SCALE 1 : 100 (cm).

N. P. V. C.
87. | 4½ HOURS

CROSS SECTION A - A

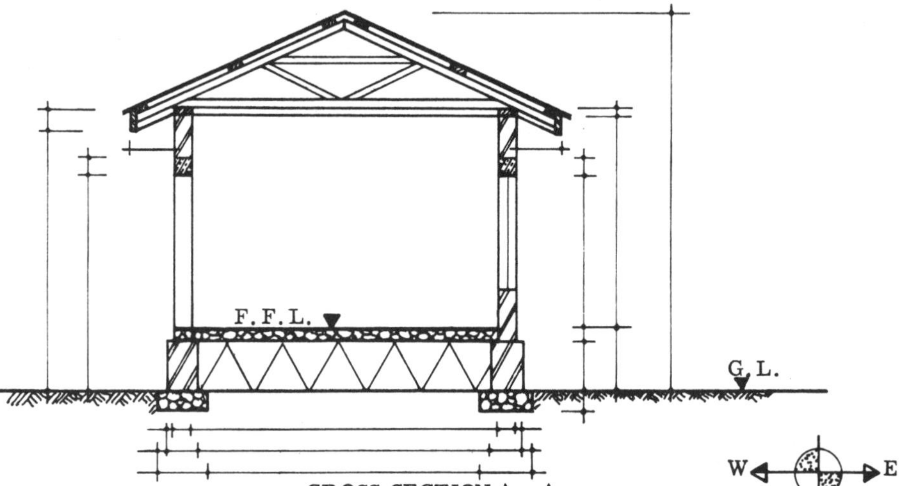

LEFT SIDE VIEW

RIGHT SIDE VIEW

TEAR THIS PAGE OUT AND GLUE IT TO PAGE 87

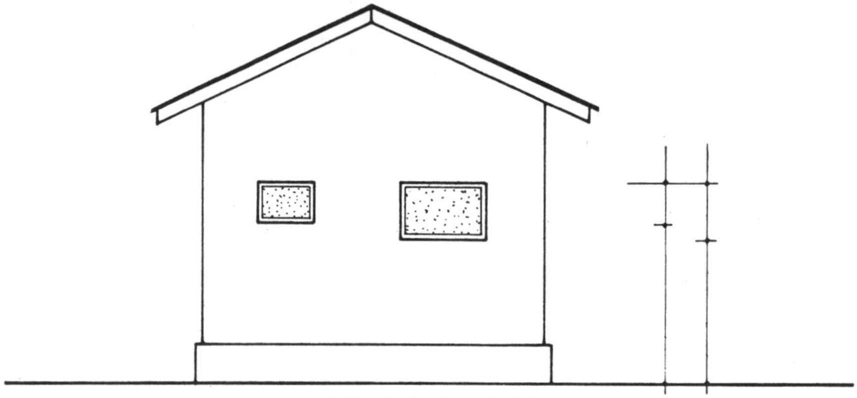

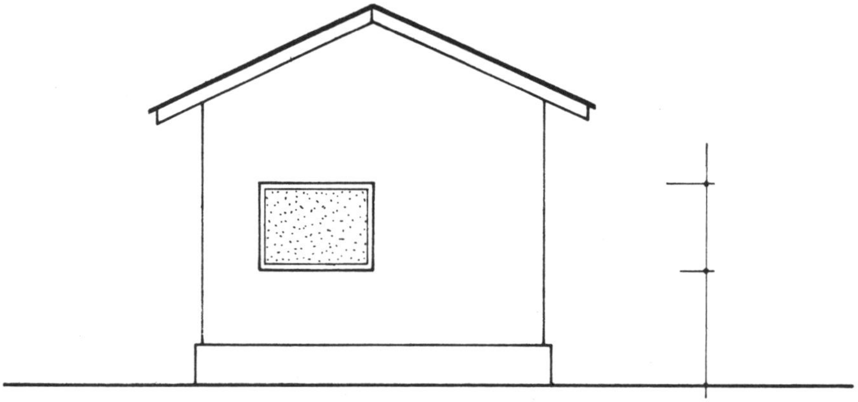

N.P.V.C.
88.

GABLE ROOF AND PARAPETTED GABLE ROOF

TOTAL LENGTH OF PARAPETTED ROOF

TOTAL LENGTH OF GABLE ROOF

CROSS SECTION

EXERCISE: NAME ALL DIMENSION LINES.
MAKE A DRAWING OF THE ABOVE ROOF DESIGNS.

N. P. V. C.
89.   3½ HOURS

PLAN FOR HOUSE WITH VERANDAH

CROSS SECTION A - A

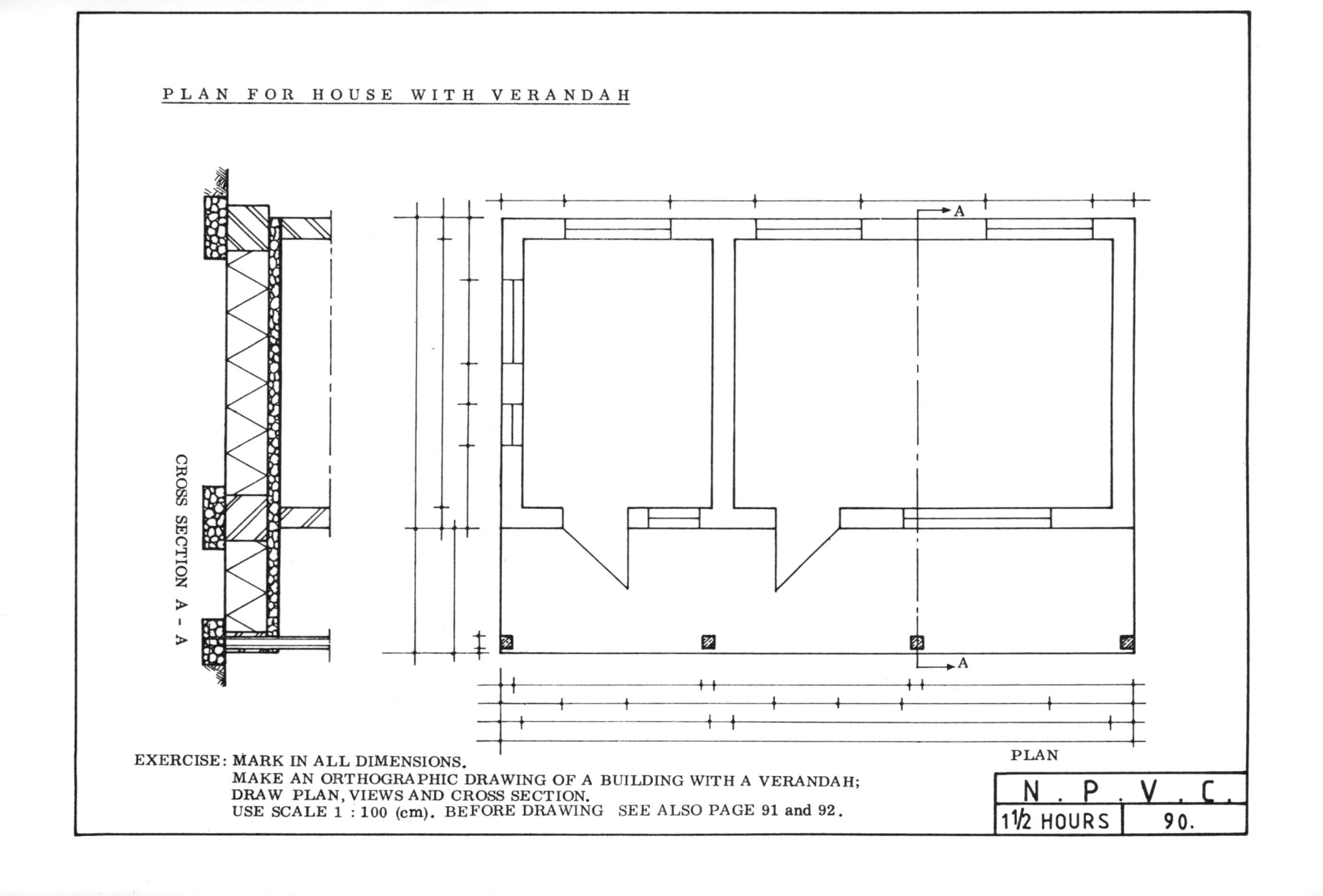

PLAN

EXERCISE: MARK IN ALL DIMENSIONS.
MAKE AN ORTHOGRAPHIC DRAWING OF A BUILDING WITH A VERANDAH;
DRAW PLAN, VIEWS AND CROSS SECTION. SEE ALSO PAGE 91 and 92.
USE SCALE 1 : 100 (cm). BEFORE DRAWING

N . P . V . C .
1½ HOURS | 90.

BUILDING WITH VERANDAH
GABLE ROOF DESIGN

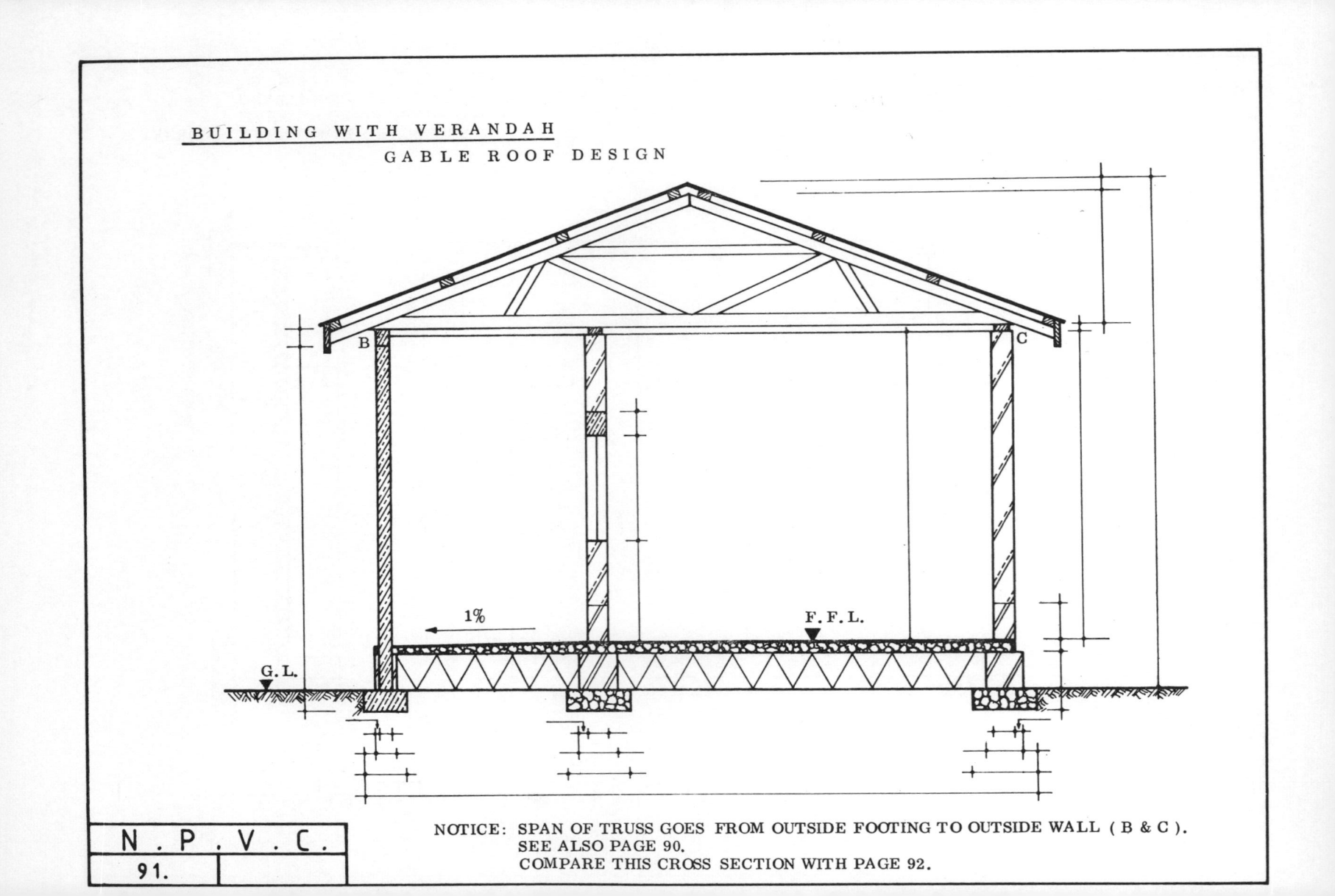

F.F.L.

G.L.

1%

B

C

NOTICE: SPAN OF TRUSS GOES FROM OUTSIDE FOOTING TO OUTSIDE WALL (B & C).
SEE ALSO PAGE 90.
COMPARE THIS CROSS SECTION WITH PAGE 92.

N.P.V.C.
91.

BUILDING WITH VERANDAH
GABLE ROOF AND OVERHANG DESIGN

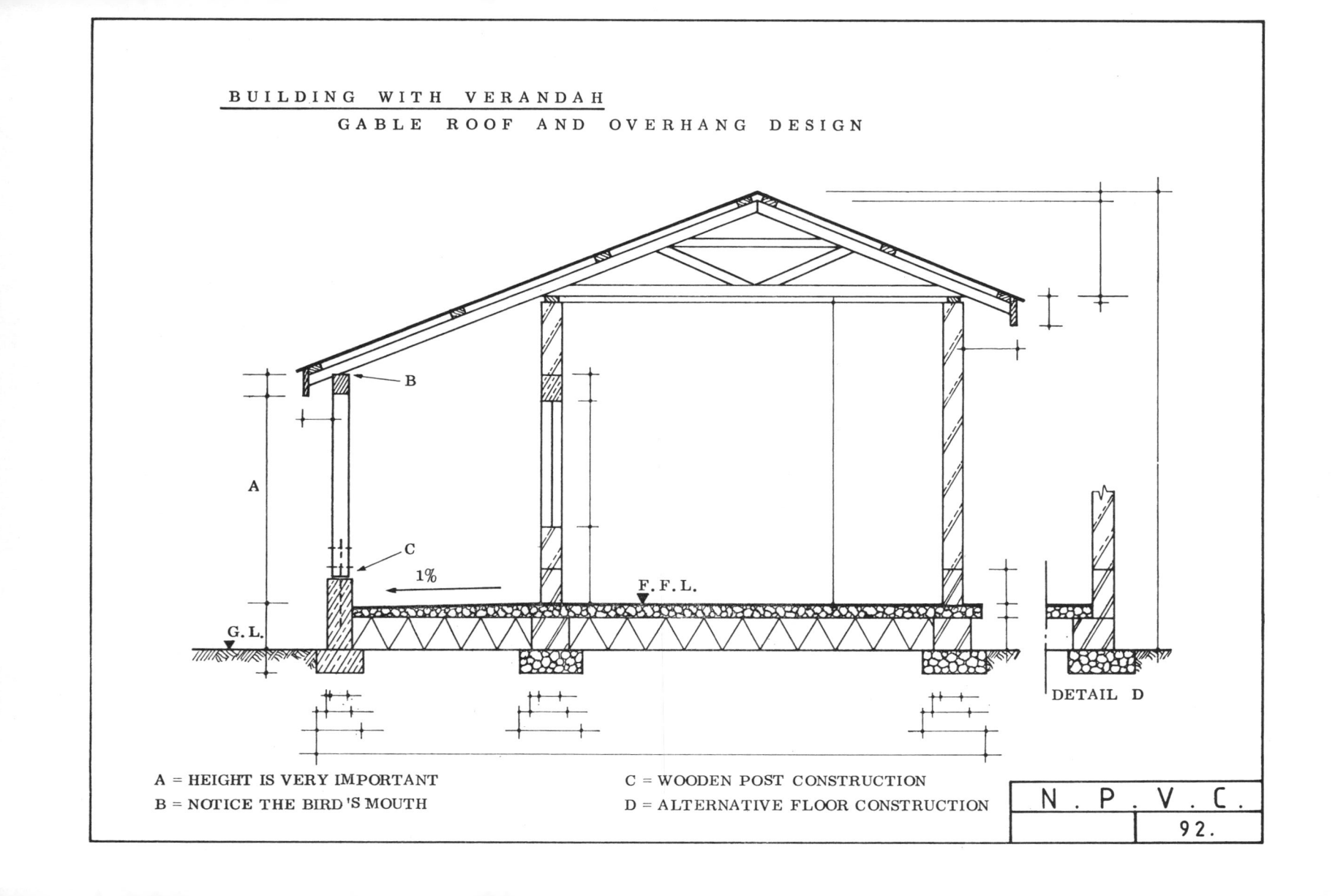

DETAIL D

F.F.L.

1%

B

C

A

G.L.

A = HEIGHT IS VERY IMPORTANT
B = NOTICE THE BIRD'S MOUTH
C = WOODEN POST CONSTRUCTION
D = ALTERNATIVE FLOOR CONSTRUCTION

N.P.V.C.

92.

FOUNDATION PLAN
BUILDING WITH VERANDAH

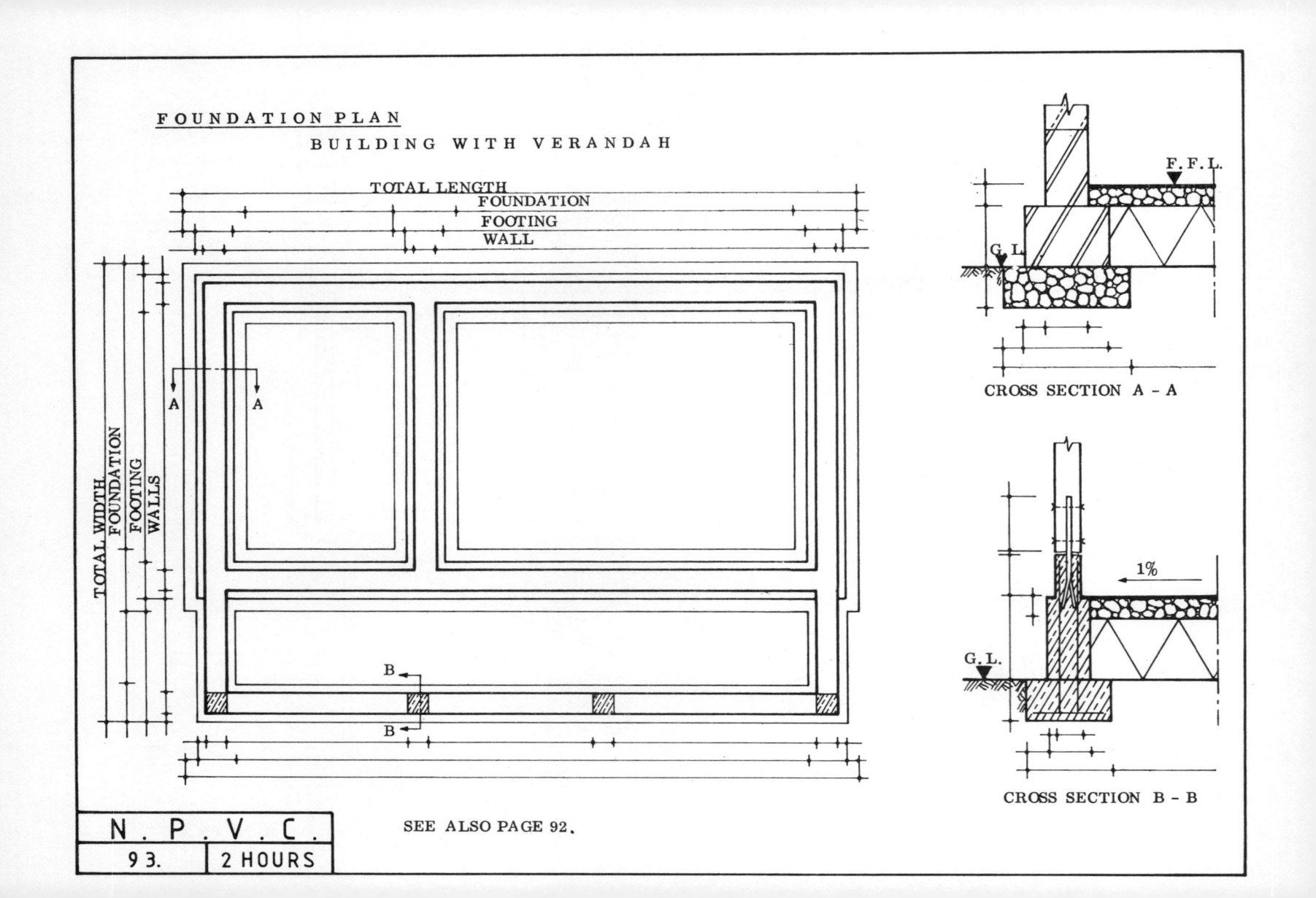

CROSS SECTION A - A

CROSS SECTION B - B

F.F.L.
G.L.
1%

TOTAL LENGTH
FOUNDATION
FOOTING
WALL

TOTAL WIDTH
FOUNDATION
FOOTING
WALLS

A — A
B — B

SEE ALSO PAGE 92.

N.P.V.C.
93.
2 HOURS

GABLE ROOF WITH OVERHANG

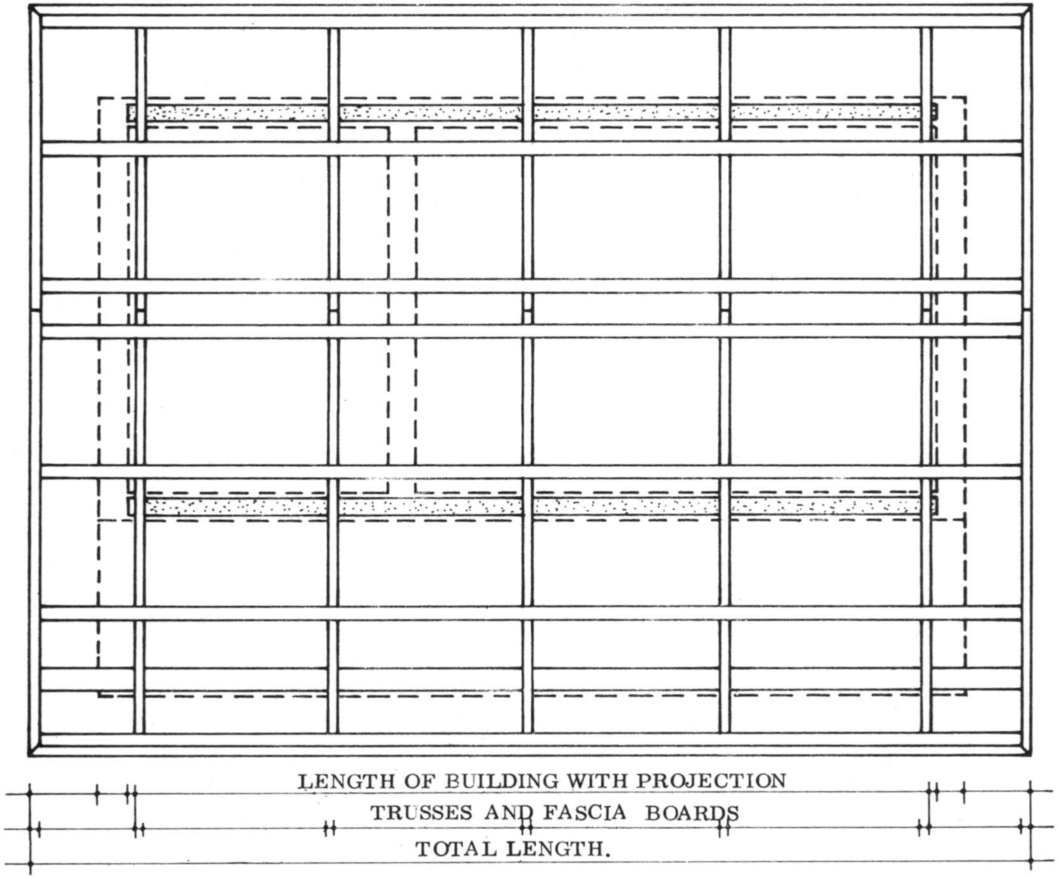

CROSS SECTION

VERANDAH | SPAN OF TRUSS

HEIGHT OF TRUSS

COVERING | PURLINS

COVERING | PURLINS

LENGTH OF BUILDING WITH PROJECTION
TRUSSES AND FASCIA BOARDS
TOTAL LENGTH.

WALL PLATE = DOTTED AREAS.
SEE ALSO PAGE 92.

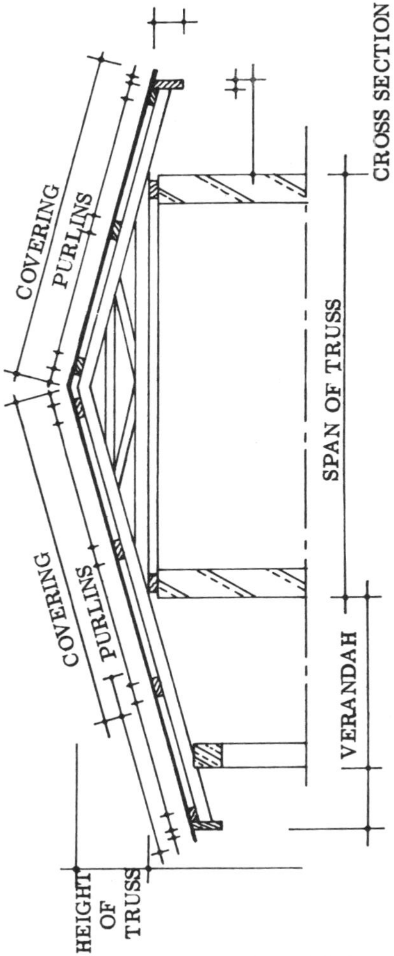

N.P.V.C.
3½ HOURS | 94.

NOTES:

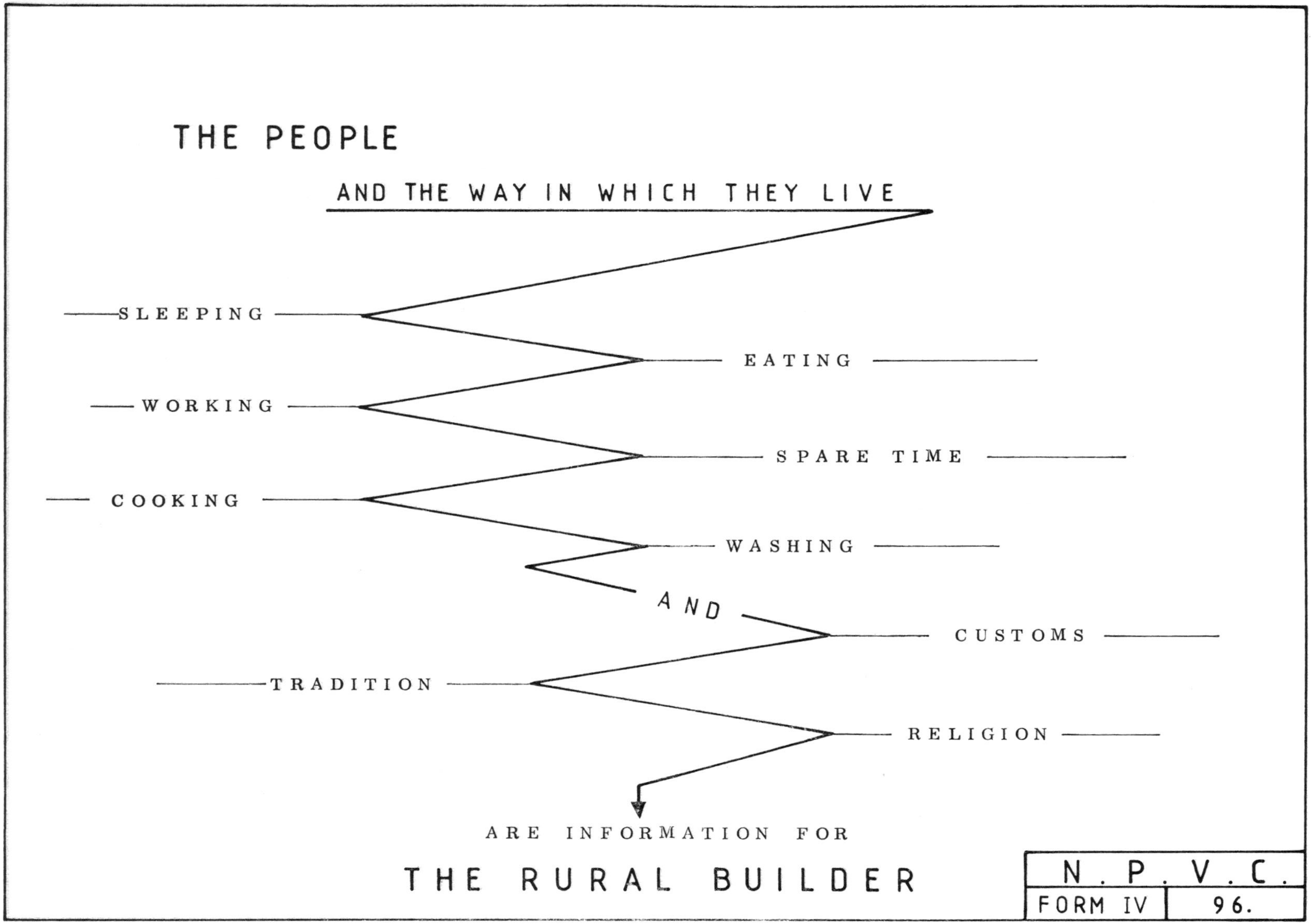

THE PEOPLE
AND THE WAY IN WHICH THEY LIVE

- SLEEPING
- EATING
- WORKING
- SPARE TIME
- COOKING
- WASHING
- AND
- TRADITION
- CUSTOMS
- RELIGION

ARE INFORMATION FOR

THE RURAL BUILDER

N.P.V.C.
FORM IV 96.

BUILDING DESIGN

You have some general technical information about building. Now you can try to make your own designs, with the aid of the basic outlines on the next pages. Keep in mind all that you know about building, and follow a policy of design:

GENERAL POLICY

Plan for the future in your design. Perhaps it may not be possible to build the entire building at once; it may be completed section by section as funds become available. The building can be planned as a whole, properly constructed piece, and then built up over the years. It is better to design what is actually needed than to design something which is not adequate and have to change it, or even abandon it when it cannot be changed into the desired structure.

Remember that a building is a lasting structure. Try to think ahead and lay out the site and the building with the future requirements in mind. Also keep in mind the points listed on page 96. Think about how people do these daily things and design around their needs.

NOTES:

DESIGN POLICY

- EFFICIENCY: This is so that the whole structure will function as it is meant to do. There should be sufficient room for all the activities and for furniture and whatever equipment is needed. Provide ventilation, privacy, protection against insects, and pay attention to water and sewage problems.

- DURABILITY: The building should withstand the stresses of its own weight, and the outside forces such as wind. It should be as protected as possible against attack from weather, dampness, insects, and the normal wear and tear of use. Keep in mind the climate and its influences, and choose your materials with some thought to their durability.

- ECONOMY: Design and erect buildings economically. Plan ahead and have the materials ready to go through with whatever section has been planned. Knowledge about materials, their quality and durability prevents much waste.

- FINISH: Finish the whole as attractively as possible. Finishes not only improve appearance, but they are usually preservatives as well. Appropriate colours make the building pleasant as well as cooler.

NOTES:

LOCATION PLAN

The illustration on the right page shows a piece of land as it looks from the ground, and how it would look if you were above it in an aeroplane looking down. The view you would see from the plane resembles the plan.

A plan tells you all the things which are on the site such as buildings, trees, roads, streams, bridges; and where all of these are located. The plan also shows the orientation of all these features with respect to directions: north, south, east and west; and how the surface of the ground slopes.

It is impossible to properly plan a building without a great deal of knowledge about the land on which it is to be built. The plan is essential to help the builder to design a building which will fit into the surroundings in the most economical and convenient way.

If a building is planned without taking into account the basic information contained in the location plan, it may turn out to be the wrong shape to fit the site. It may be more costly to build because it is not placed correctly along the "contours" of the land. It may be uncomfortable to live in if it is facing into the hot afternoon sun. It may be hard to reach if there is not a good entrance way from the road.

The following pages contain some information about the factors which need to be considered about the site when you are planning a building. These include the measurements of the site and plot, the building regulations and restrictions, the direction of the breeze and of storms, the slope of the site, the direction of the sun's rays at different times of day: all of these need to be considered when planning so that the building will be comfortable to live in and economical to build.

IF YOU LOOK DOWN FROM AN AEROPLANE:

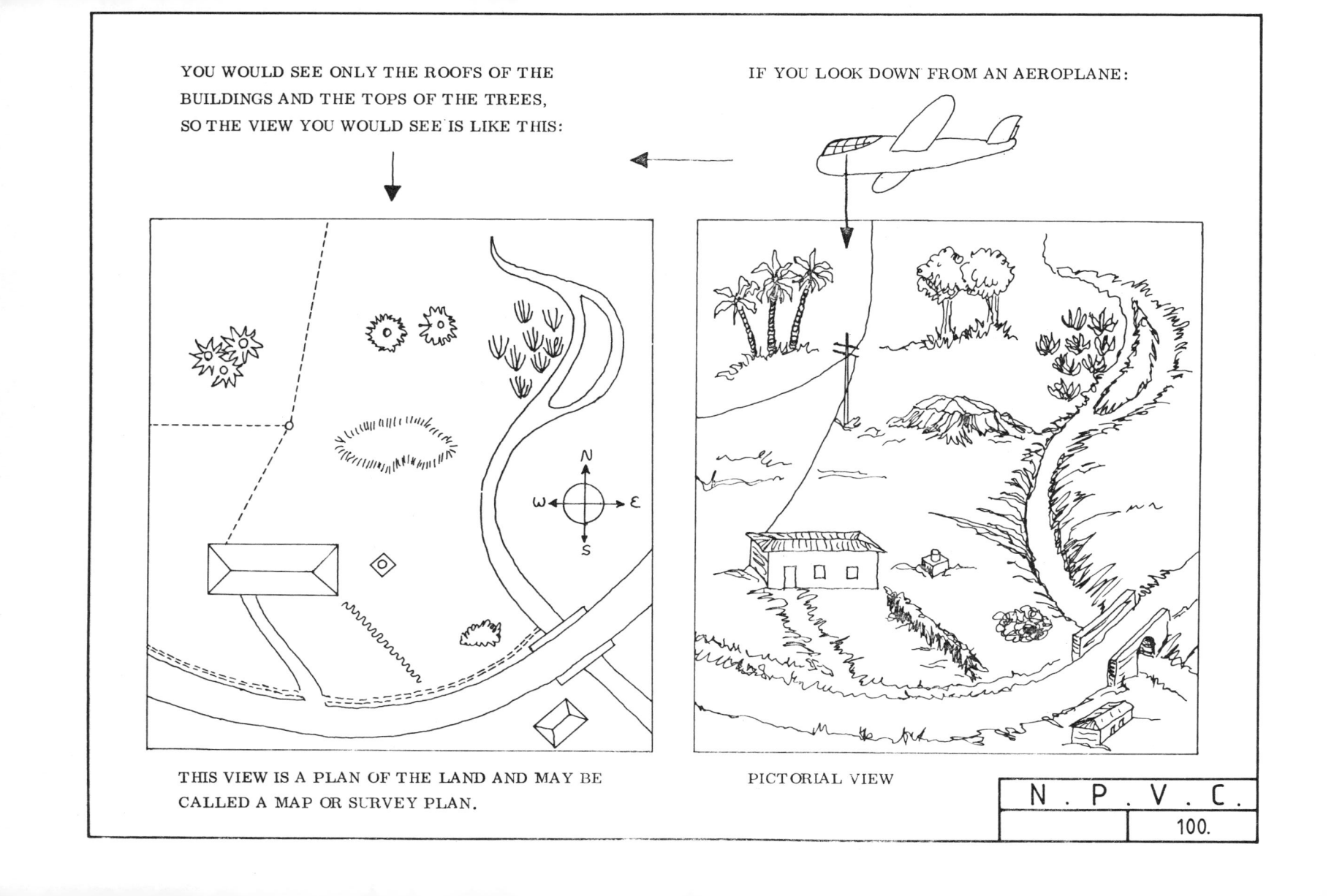

YOU WOULD SEE ONLY THE ROOFS OF THE
BUILDINGS AND THE TOPS OF THE TREES,
SO THE VIEW YOU WOULD SEE IS LIKE THIS:

PICTORIAL VIEW

THIS VIEW IS A PLAN OF THE LAND AND MAY BE
CALLED A MAP OR SURVEY PLAN.

N.	P.	V.	C.
			100.

MEASUREMENTS
POINT TO POINT

"BOUNDARY STONE". THE ARROW IS FACING THE NORTH. THE NEXT STONE IS 73 METRES AWAY.

SITE PLANS HAVE THE LENGTHS WRITTEN ALONG THE BOUNDARIES SO THAT THE EXACT LOCATION OF THE PLOT IS KNOWN.

BOUNDARY LINE

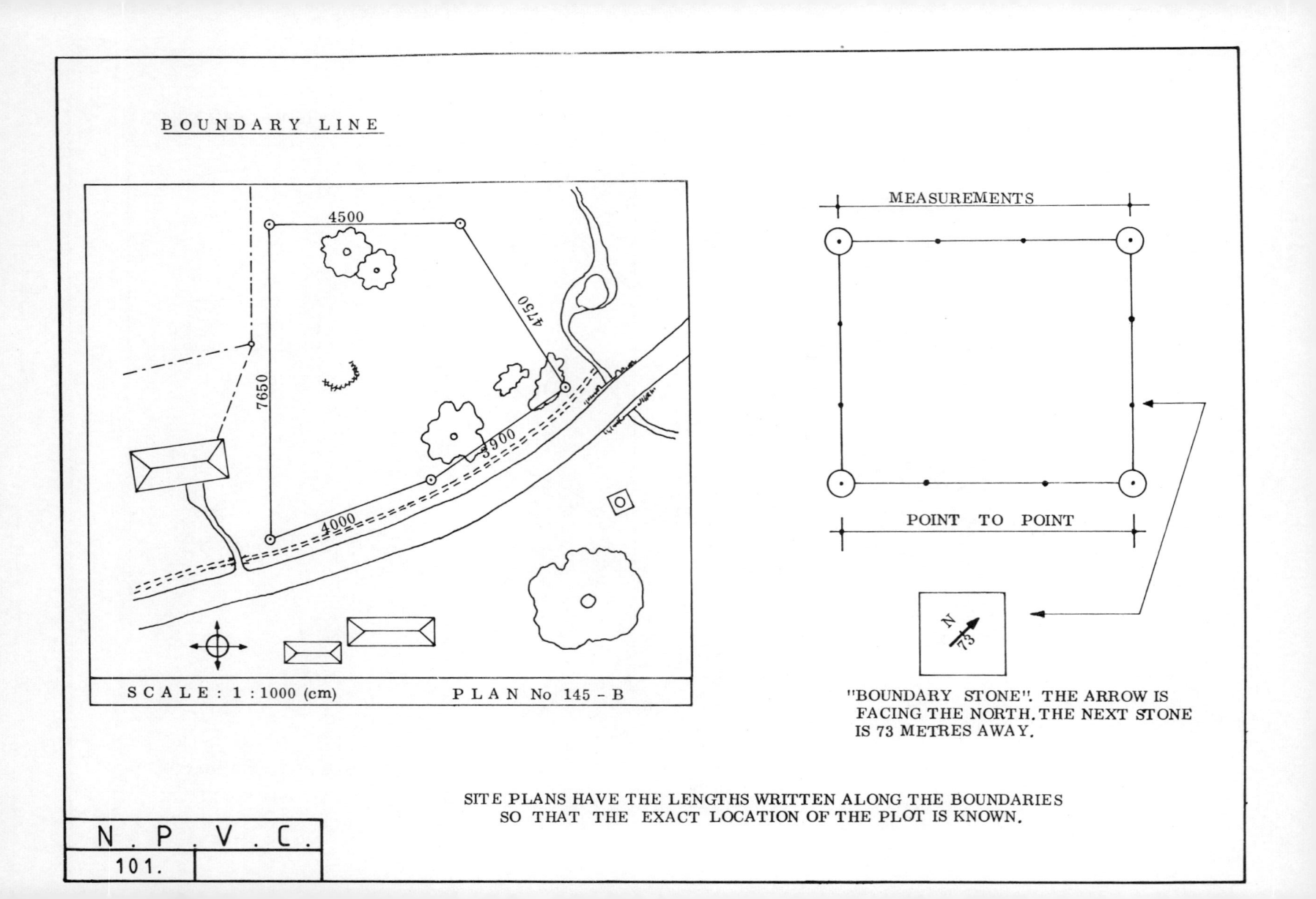

SCALE : 1 : 1000 (cm)   PLAN No 145 - B

N. P. V. C.
101.

BUILDING LINE

BUILDINGS

ROAD

BUILDINGS

IN PLANNED TOWNS NO BUILDING IS ALLOWED TO REACH TO THE EDGE OF THE ROAD: A BUILDING HAS TO BE A CERTAIN DISTANCE AWAY FROM THE ROAD TO ALLOW FOR FUTURE ROAD WIDENING AND SO THAT THERE IS ENOUGH SPACE BETWEEN BUILDINGS.

SITE COVERAGE

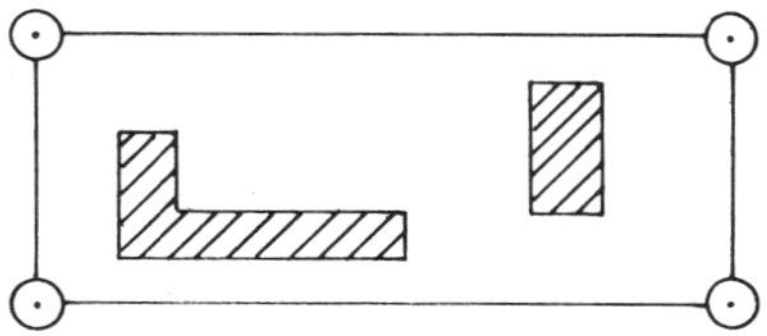

IN SOME TOWNS THERE ARE RULES ABOUT HOW MUCH LAND CAN BE BUILT OVER: FOR SO MANY SQUARE METRES OF LAND, THE BUILDINGS CAN TAKE UP ONLY A CERTAIN AMOUNT OF THE SPACE. IN ADDITION, THE BUILDINGS USUALLY HAVE TO BE A CERTAIN DISTANCE AWAY FROM THE PLOT BOUNDARIES. THIS IS TO MAKE SURE THAT THE BUILDINGS WILL HAVE PLENTY OF LIGHT AND AIR SO THEY ARE HEALTHY TO LIVE IN.

BREEZE

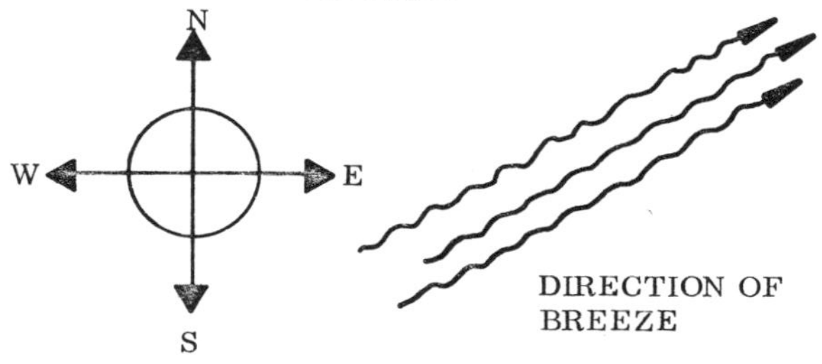

DIRECTION OF BREEZE

IN HOT WET CLIMATES, IT IS IMPORTANT TO HAVE A BREEZE BLOWING THROUGH THE BUILDING SO THAT THE PEOPLE INSIDE ARE COOLER.

THE DIRECTION OF THE PREVAILING BREEZE (THE USUAL DIRECTION THE BREEZE COMES FROM) IS MARKED ON THE PLANS.

STORMS

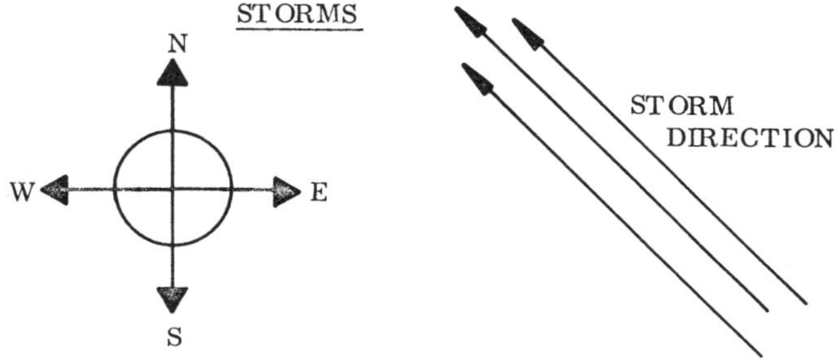

STORM DIRECTION

IN AREAS WHICH GET BAD STORMS, THE RURAL BUILDER WILL PLAN THE BUILDING SO THAT THE OPENINGS WHICH FACE THE STORM ARE PROTECTED.

N.P.V.C.

102.

BUILDING ALONG THE CONTOURS

600
550
525
500
475
450
425
400
375
350
325
300

THIS IS CHEAP

COSTLY
100 cm EXTRA FOOTING!

SLOPE

WHERE THE GROUND SLOPES STEEPLY THE RURAL BUILDER NEEDS TO SHOW THE CONTOUR LINES ON THE PLAN. THESE ARE DRAWN AT EVERY 50 cm OF HEIGHT DIFFERENCE, OR LESS IF NECESSARY.

N.P.V.C.
103.

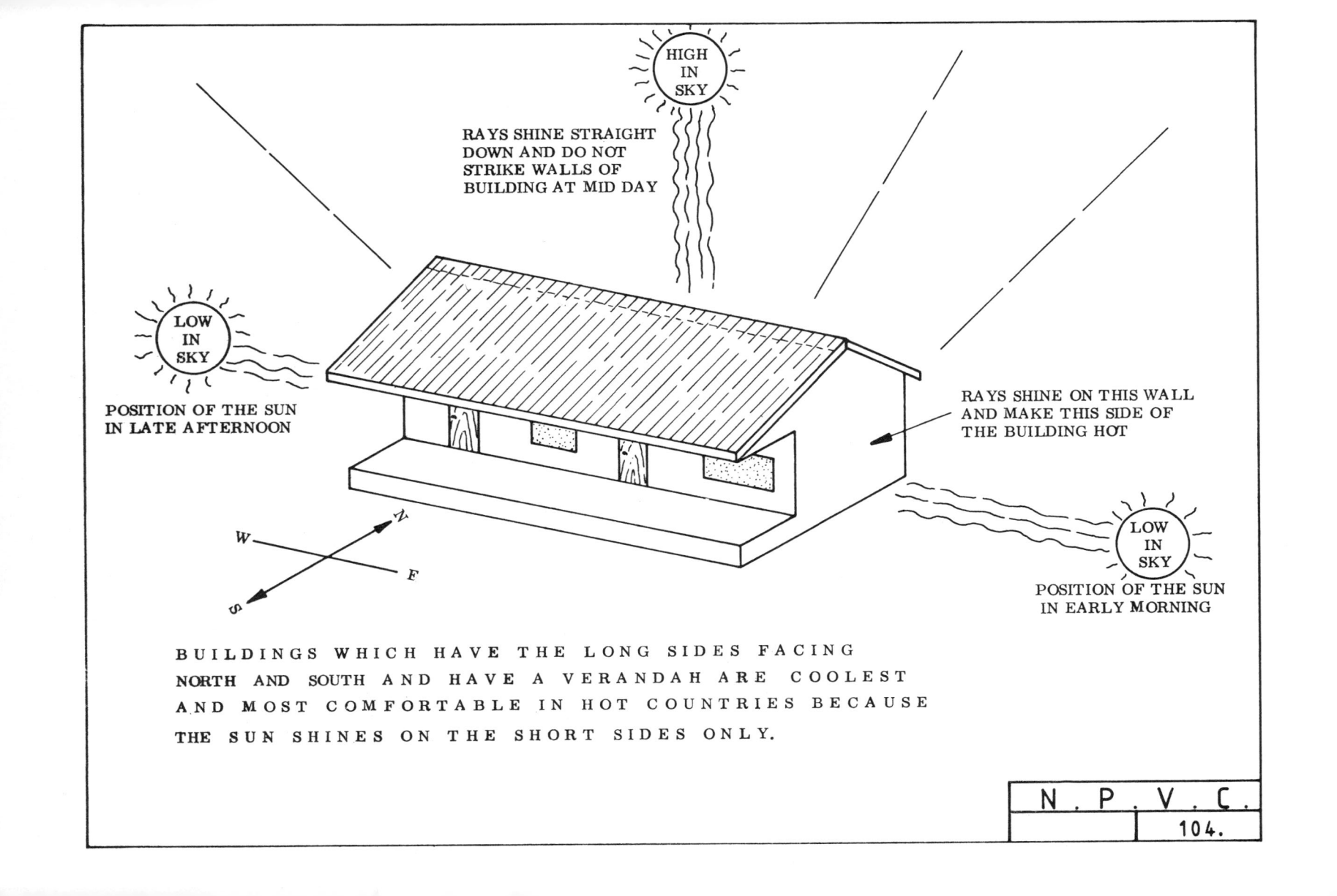

POSITION OF THE SUN IN EARLY MORNING

LOW IN SKY

RAYS SHINE ON THIS WALL AND MAKE THIS SIDE OF THE BUILDING HOT

HIGH IN SKY

RAYS SHINE STRAIGHT DOWN AND DO NOT STRIKE WALLS OF BUILDING AT MID DAY

LOW IN SKY

POSITION OF THE SUN IN LATE AFTERNOON

BUILDINGS WHICH HAVE THE LONG SIDES FACING NORTH AND SOUTH AND HAVE A VERANDAH ARE COOLEST AND MOST COMFORTABLE IN HOT COUNTRIES BECAUSE THE SUN SHINES ON THE SHORT SIDES ONLY.

DESIGNING BUILDINGS TO WITHSTAND STRONG WINDS

Many buildings are not strong enough to resist the forces of very strong winds. They may be destroyed and the people inside can be injured or even killed. The guidelines here aim to help the Rural Builder to design and construct buildings so that such occurrences are reduced in the future. The point of these guidelines is to reduce the force of the wind against the building, and to make the building more resistant to wind forces. The following points are illustrated on the right.

A - Take advantage of natural windbreaks such as trees or hedges when deciding on the site for the building. Such a location can reduce the force of the wind.

B - Sites on hills and near hilltops can have much higher windspeeds.

C - Valleys can funnel winds and create higher windspeeds.

D - The pitch of the roof is very important. This should be between 15 and 20 degrees.

E - A hip roof resists wind forces better than a gable roof.

F - Avoid making large overhangs, even if they are supported by columns. Locate verandahs away from the direction of the strongest winds.

G - A parapet around the roof helps to reduce the wind force along the roof edges.

H - Avoid making large openings such as doors or windows near the roof line or near the corners of walls. These tend to weaken the structure if they are located where the loads are greatest.

I - Make sure that every part of the building is secured: the roof parts to each other, the roof itself to the walls, the walls to the other walls, the walls to the floors, the floor to the foundations. The foundations should rest on firm soil if possible.

- REMEMBER: Whatever the form of roof construction, the parts of the roof must be securely tied together. Anchor the whole structure to the building. Ignoring this precaution means that the roof will almost certainly be damaged in any strong wind.

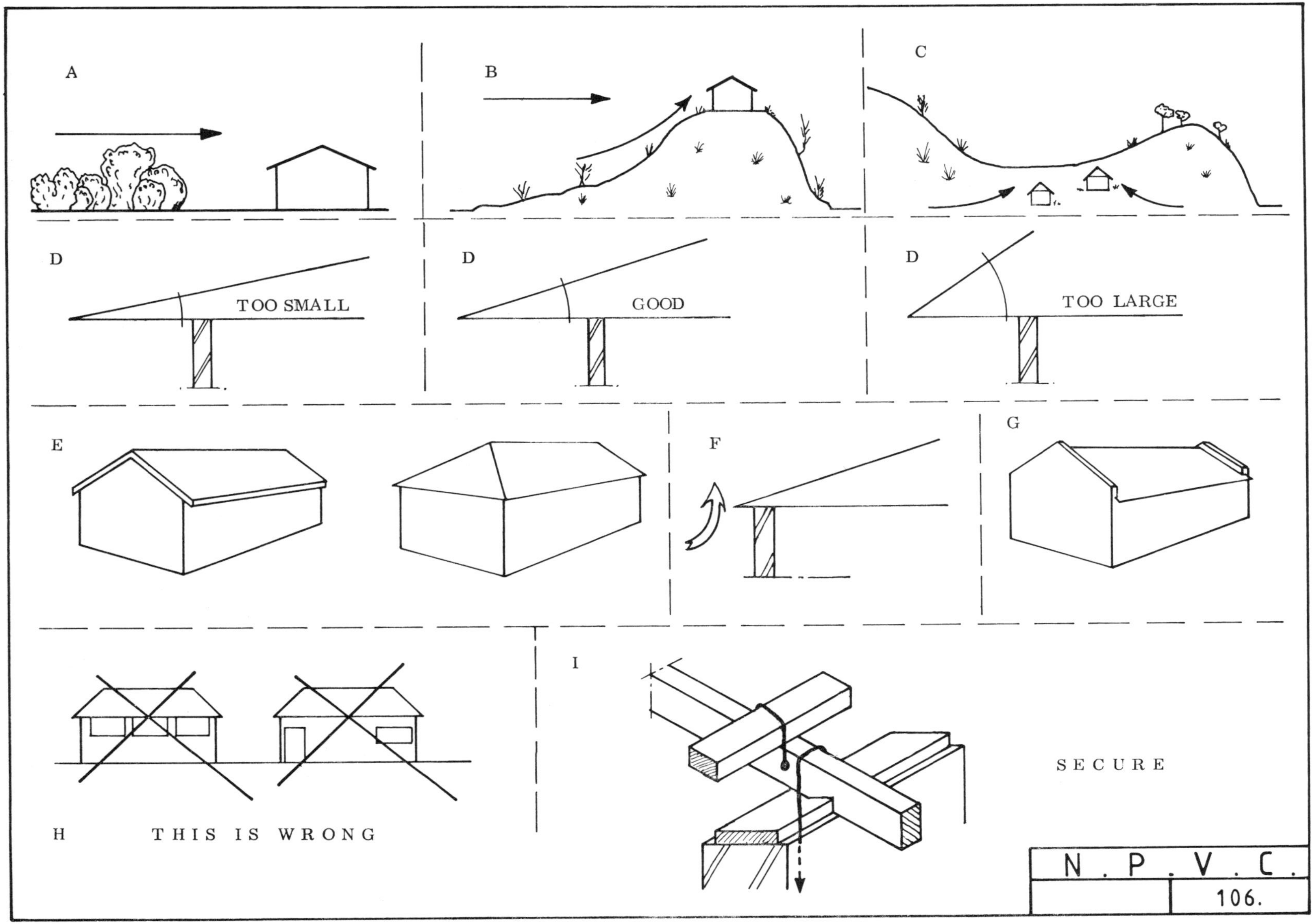

A

B

C

D TOO SMALL

D GOOD

D TOO LARGE

E

F

G

H THIS IS WRONG

I SECURE

N.P.V.C.
106.

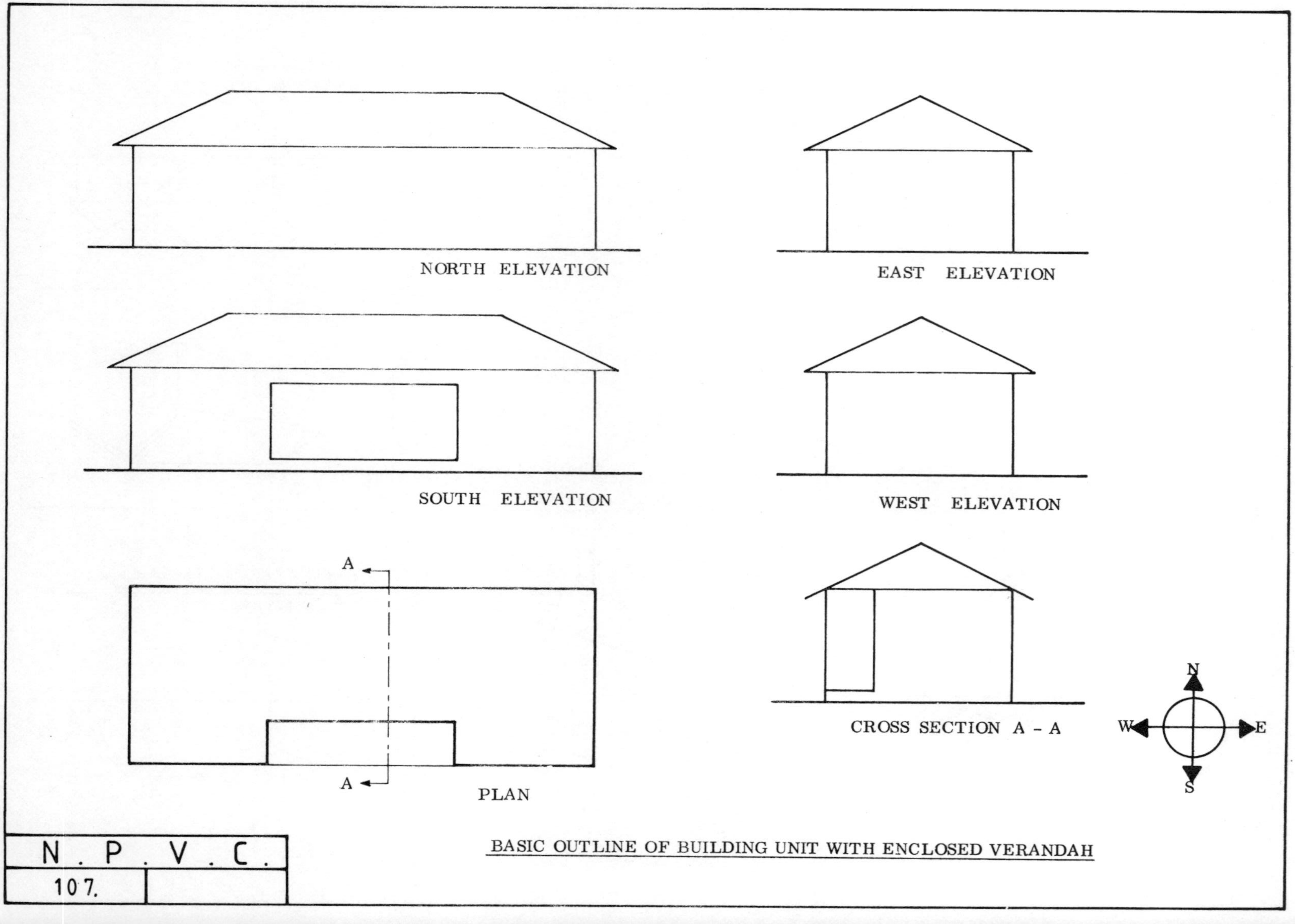

EAST ELEVATION

WEST ELEVATION

CROSS SECTION A - A

NORTH ELEVATION

SOUTH ELEVATION

PLAN

BASIC OUTLINE OF BUILDING UNIT WITH ENCLOSED VERANDAH

N. P. V. C.

107.

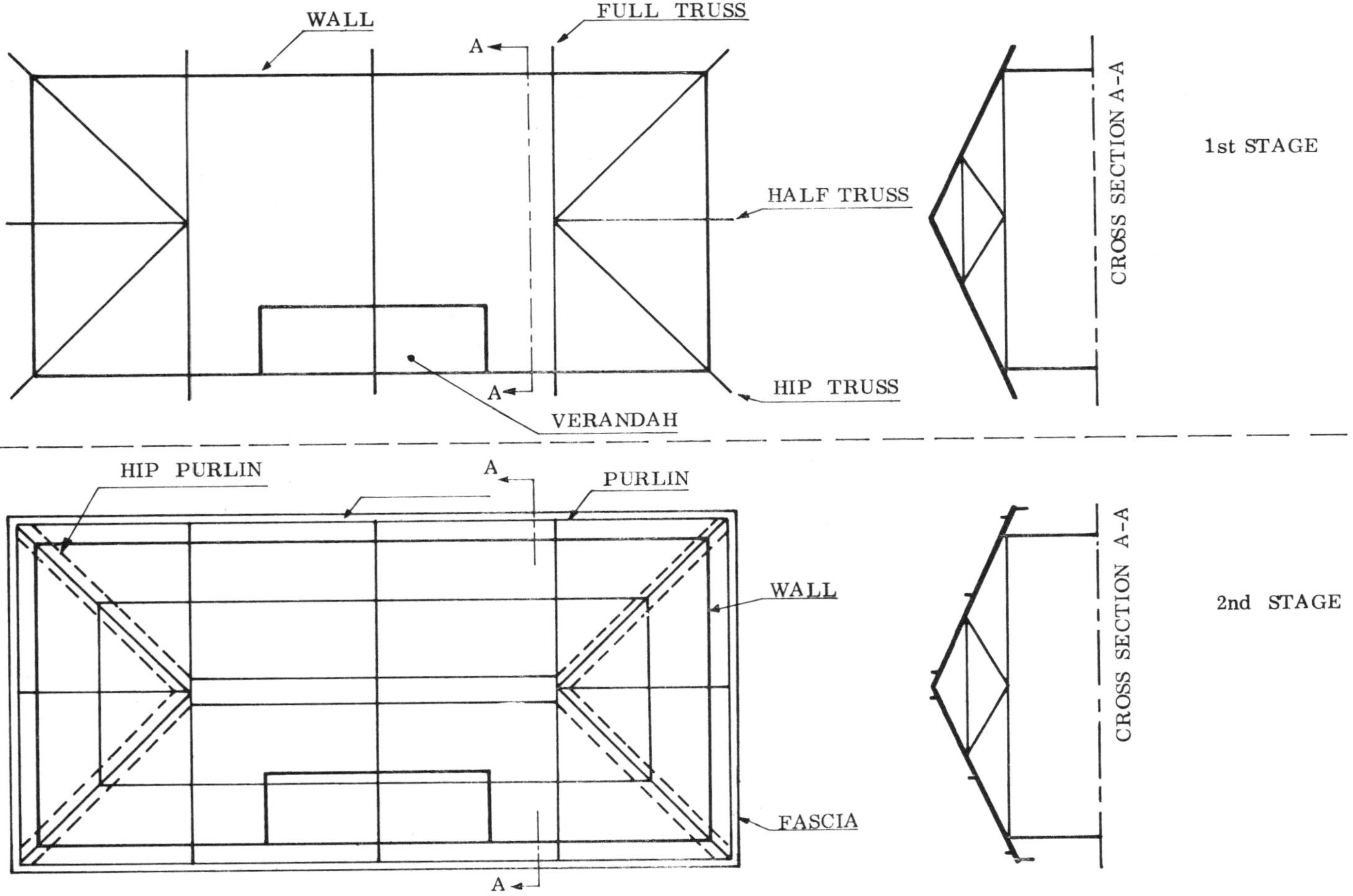

BASIC OUTLINE FOR HIP ROOF CONSTRUCTION

1st STAGE

CROSS SECTION A-A

FULL TRUSS
HALF TRUSS
HIP TRUSS
WALL
VERANDAH

2nd STAGE

CROSS SECTION A-A

WALL
FASCIA
PURLIN
HIP PURLIN

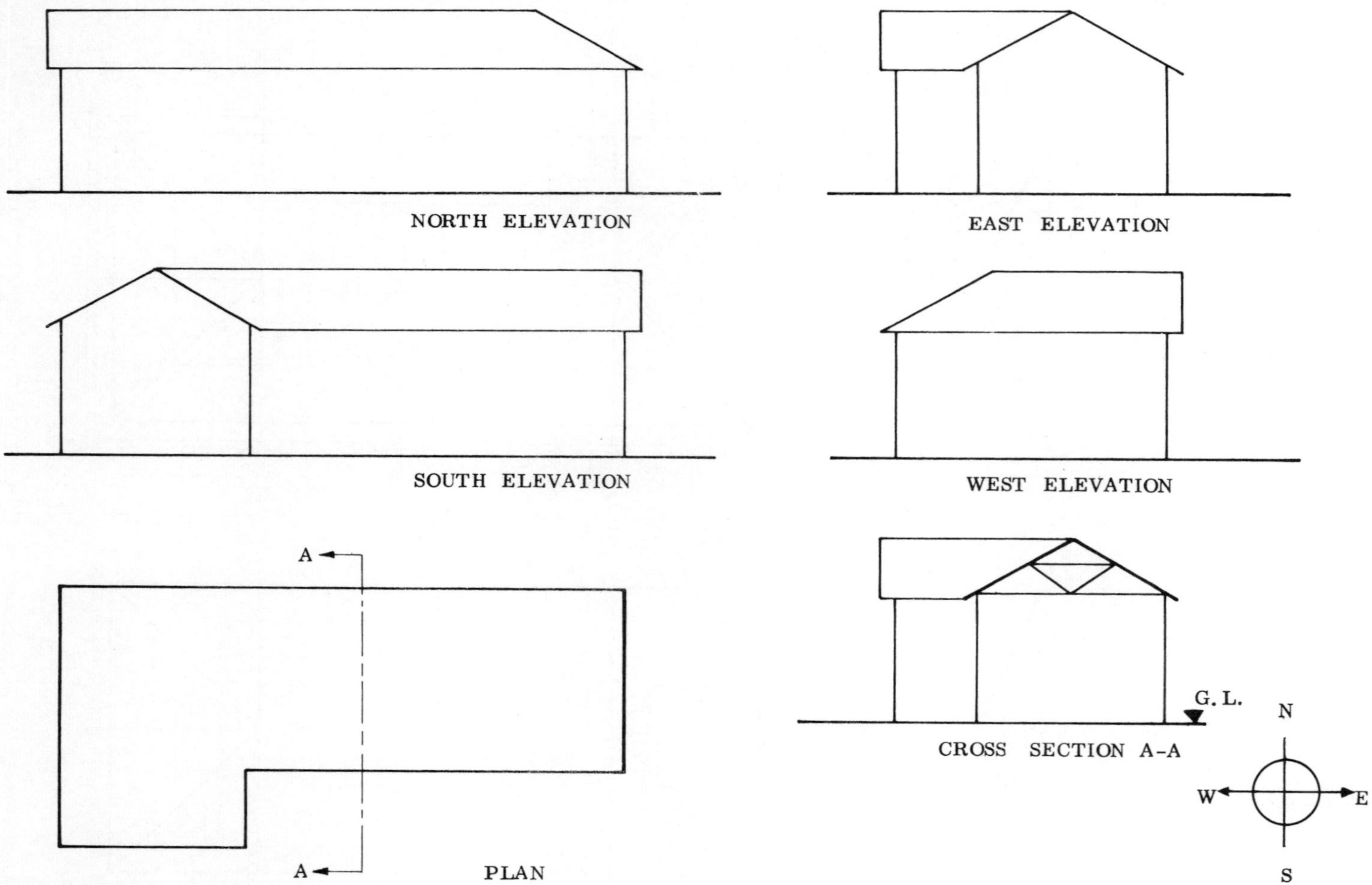

EAST ELEVATION

WEST ELEVATION

CROSS SECTION A-A

NORTH ELEVATION

SOUTH ELEVATION

PLAN

BASIC OUTLINE "L" SHAPED BUILDING UNIT

N.P.V.C.

109.

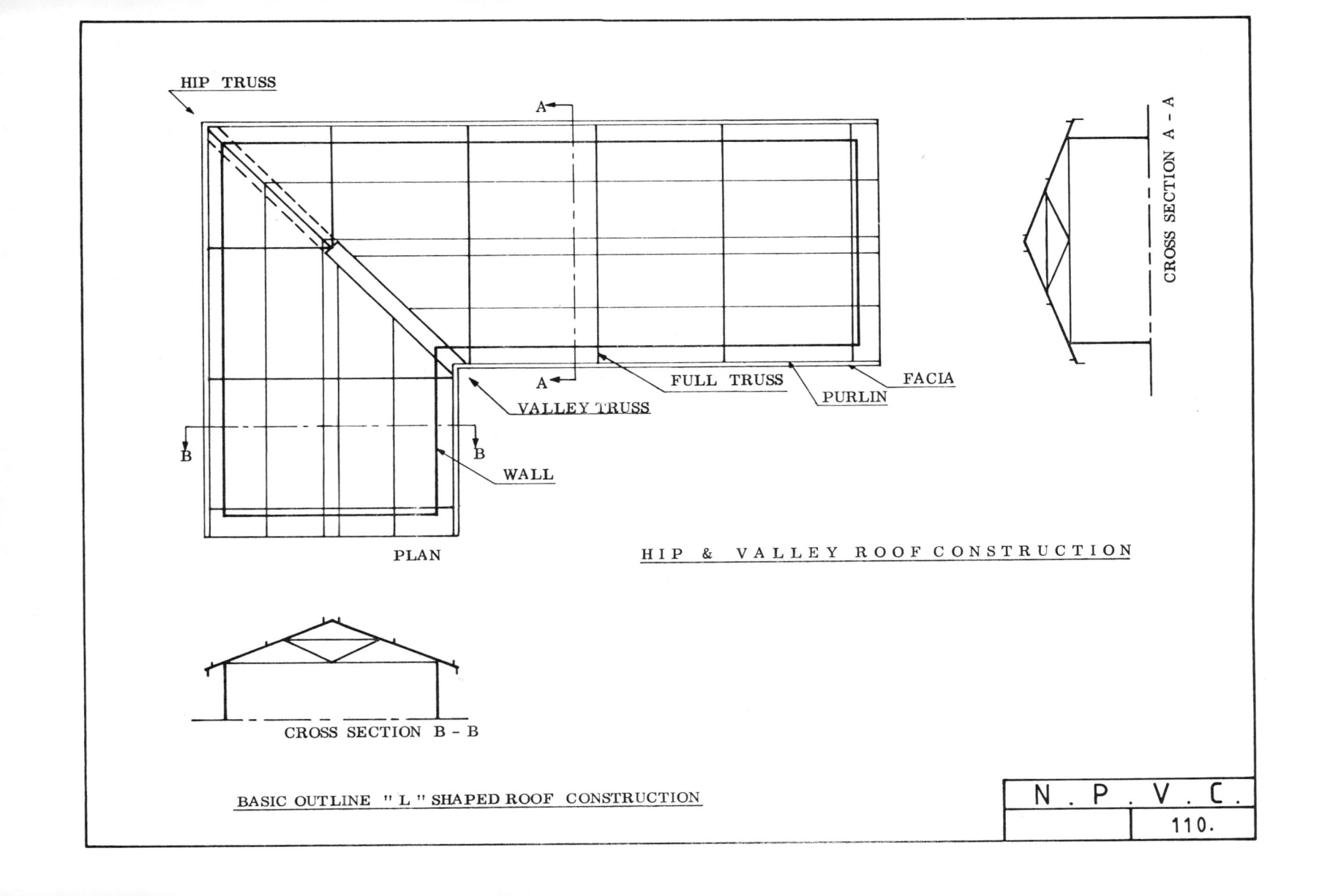

CROSS SECTION A - A

FACIA
PURLIN
FULL TRUSS
VALLEY TRUSS
A
A
WALL
B
B
PLAN
HIP TRUSS

CROSS SECTION B - B

HIP & VALLEY ROOF CONSTRUCTION

BASIC OUTLINE " L " SHAPED ROOF CONSTRUCTION

N . P . V . C.

110.

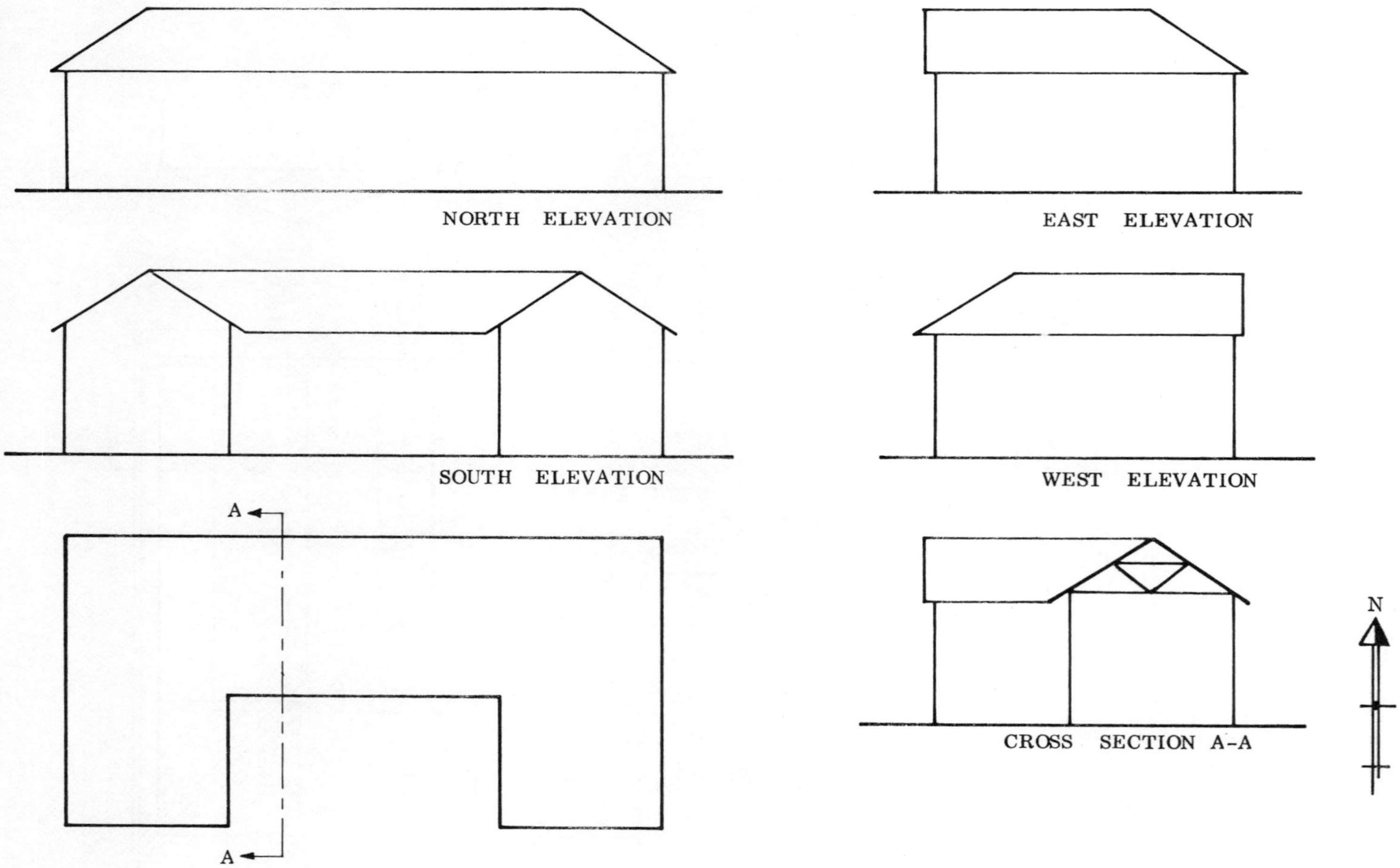

EAST ELEVATION

WEST ELEVATION

CROSS SECTION A-A

NORTH ELEVATION

SOUTH ELEVATION

N

A A

BASIC OUTLINE "U" SHAPED BUILDING UNIT

N.P.V.C.

111.

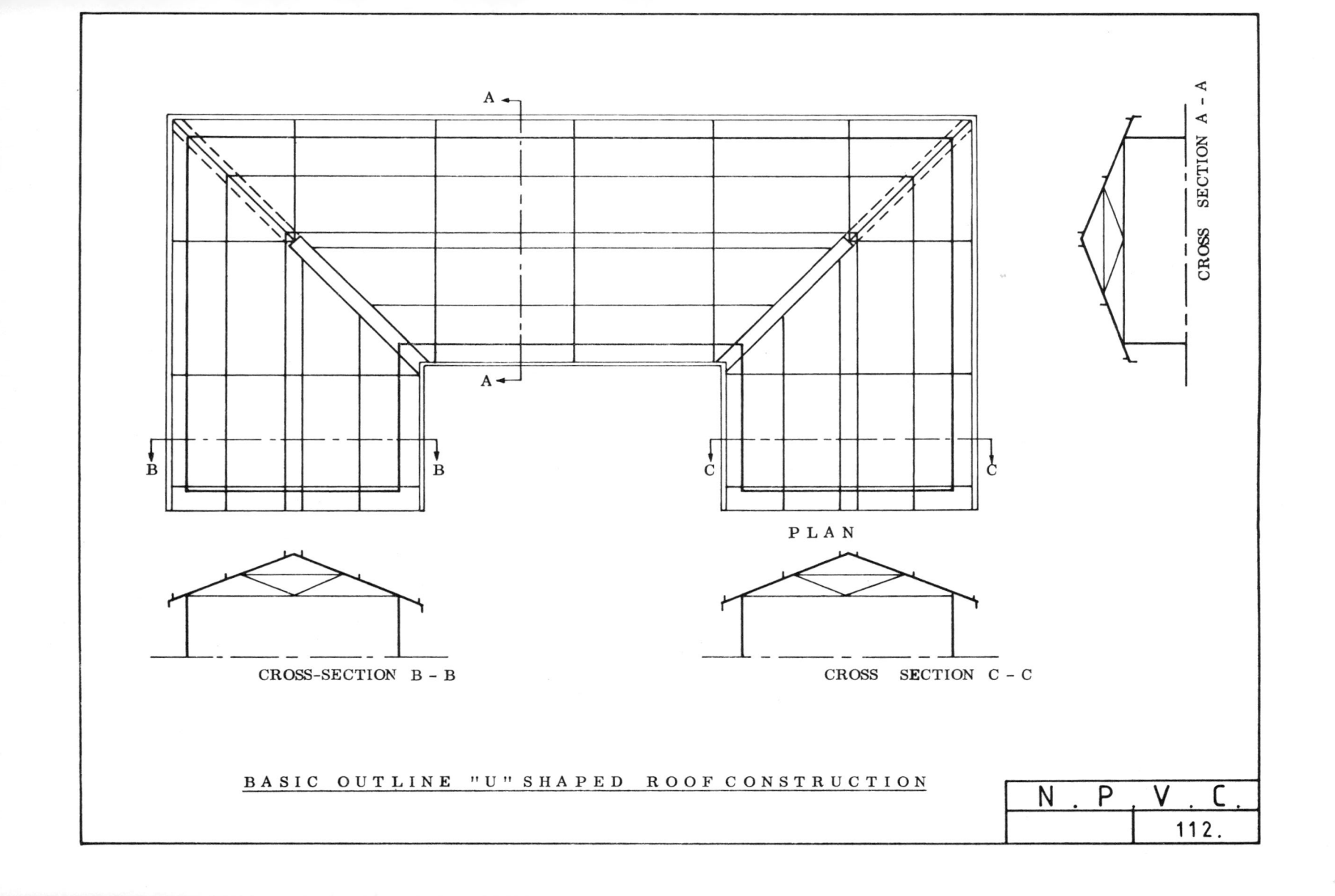

CROSS SECTION A - A

PLAN

CROSS SECTION C - C

CROSS-SECTION B - B

BASIC OUTLINE "U" SHAPED ROOF CONSTRUCTION

N.P.V.C.

112.

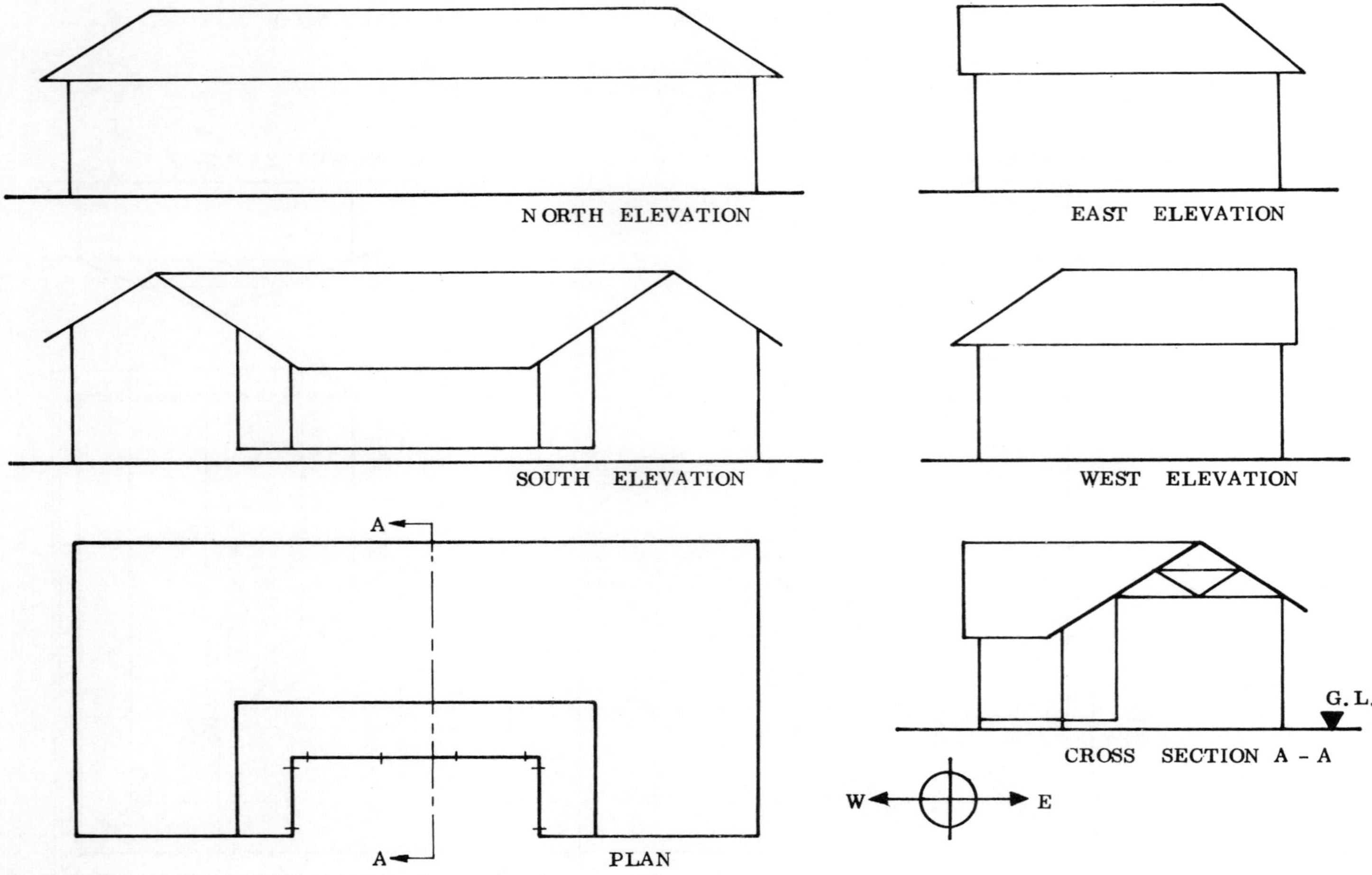

EAST ELEVATION

WEST ELEVATION

CROSS SECTION A - A

NORTH ELEVATION

SOUTH ELEVATION

PLAN

BASIC OUTLINE "U" SHAPED BUILDING UNIT
WITH RAFTER EXTENDING OVER VERANDAH

N. P. V. C.

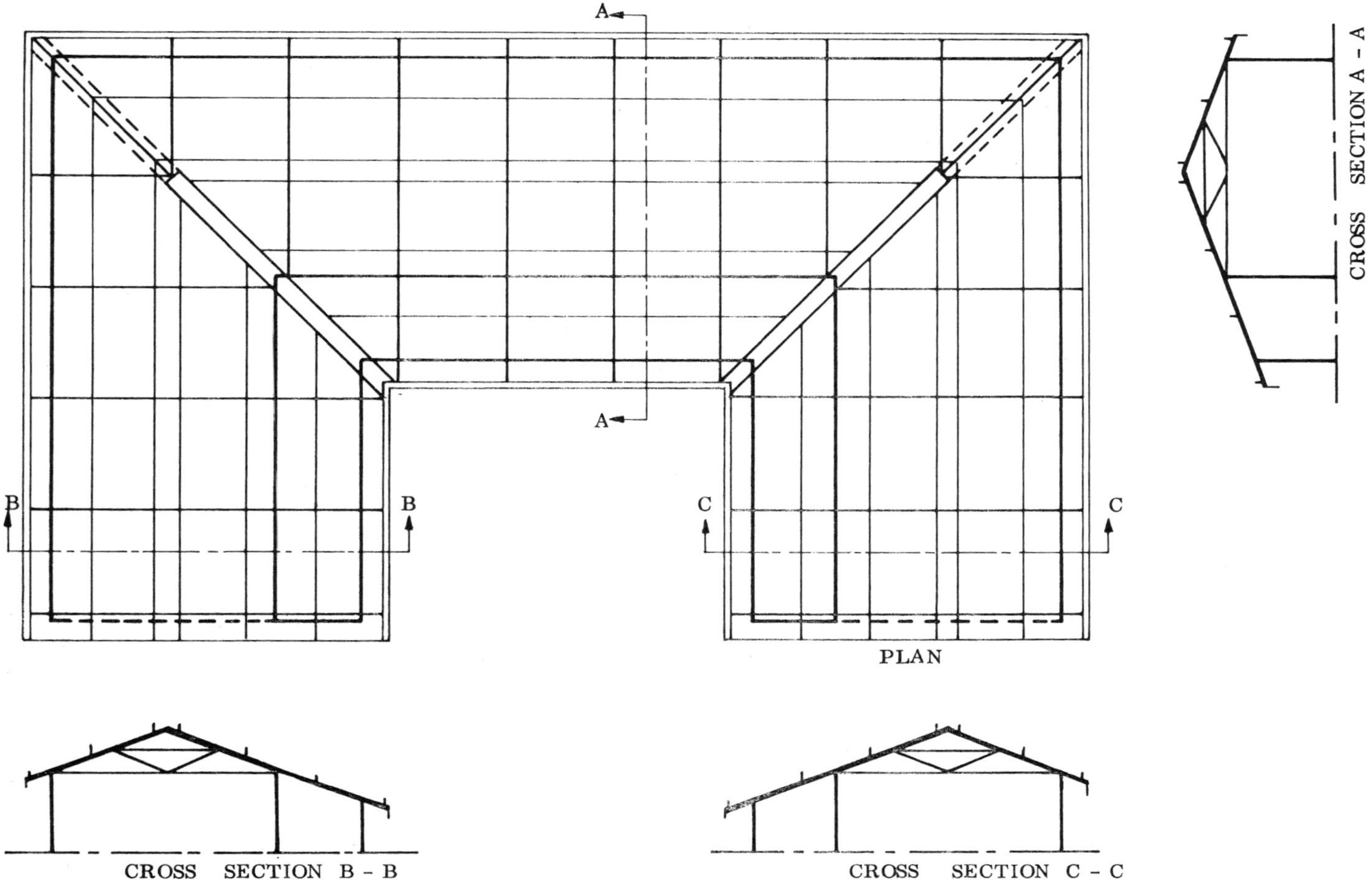

CROSS SECTION A - A

PLAN

CROSS SECTION C - C

CROSS SECTION B - B

BASIC OUTLINE "U" SHAPED ROOF WITH RAFTER
EXTENDING OVER VERANDAH

N.P.V.C.

114.

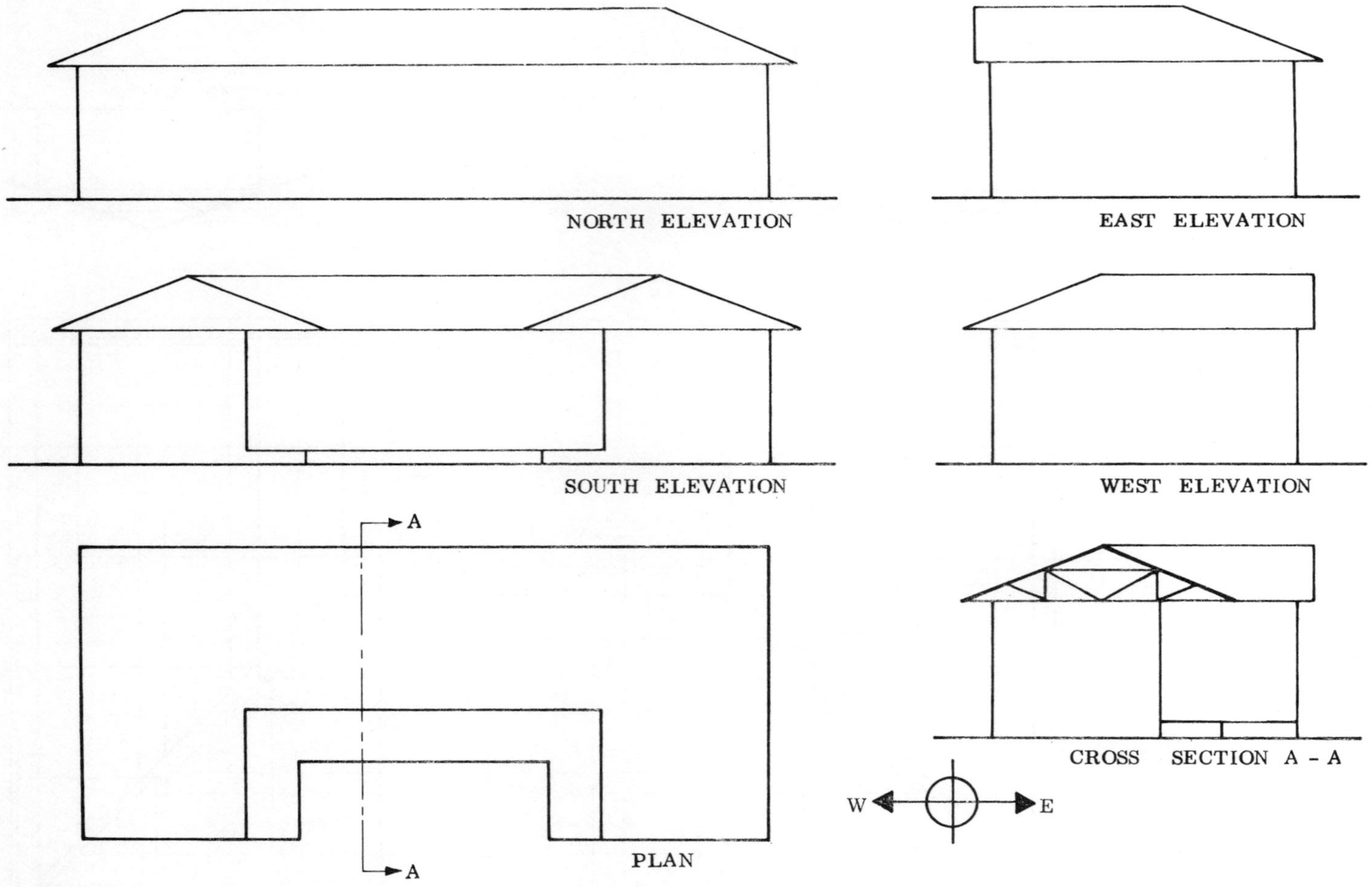

EAST ELEVATION

WEST ELEVATION

CROSS SECTION A - A

NORTH ELEVATION

SOUTH ELEVATION

PLAN

BASIC OUTLINE "U" SHAPED BUILDING UNIT
WITH TIE BEAM EXTENDING OVER VERANDAH

N . P . V . C .
115.

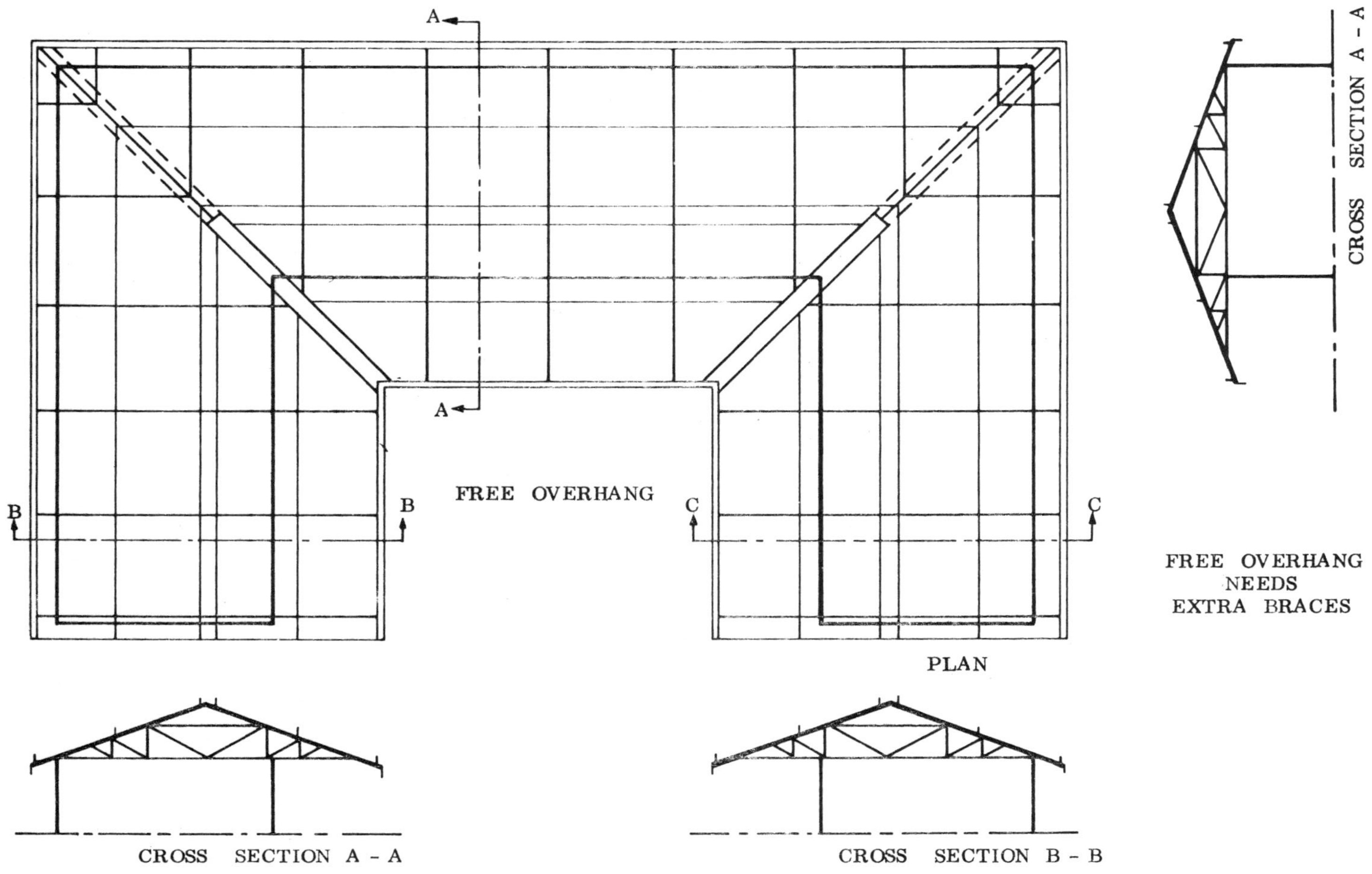

CROSS SECTION A - A

FREE OVERHANG
NEEDS
EXTRA BRACES

PLAN

FREE OVERHANG

CROSS SECTION B - B

CROSS SECTION A - A

BASIC OUTLINE "U" SHAPED ROOF WITH TIE BEAM
EXTENDING OVER VERANDAH

N. P. V. C.

116.

d = LENGTH OF BLOCK

COURSE 2 - 4 ETC.

LENGTH OF BLOCK

JOINT

TOP VIEW

COURSE 1 - 3 ETC.

TOP VIEW

CIRCULAR WORK

SIDE VIEW

G.L.

N.P.V.C.

117.

GRAIN SILO

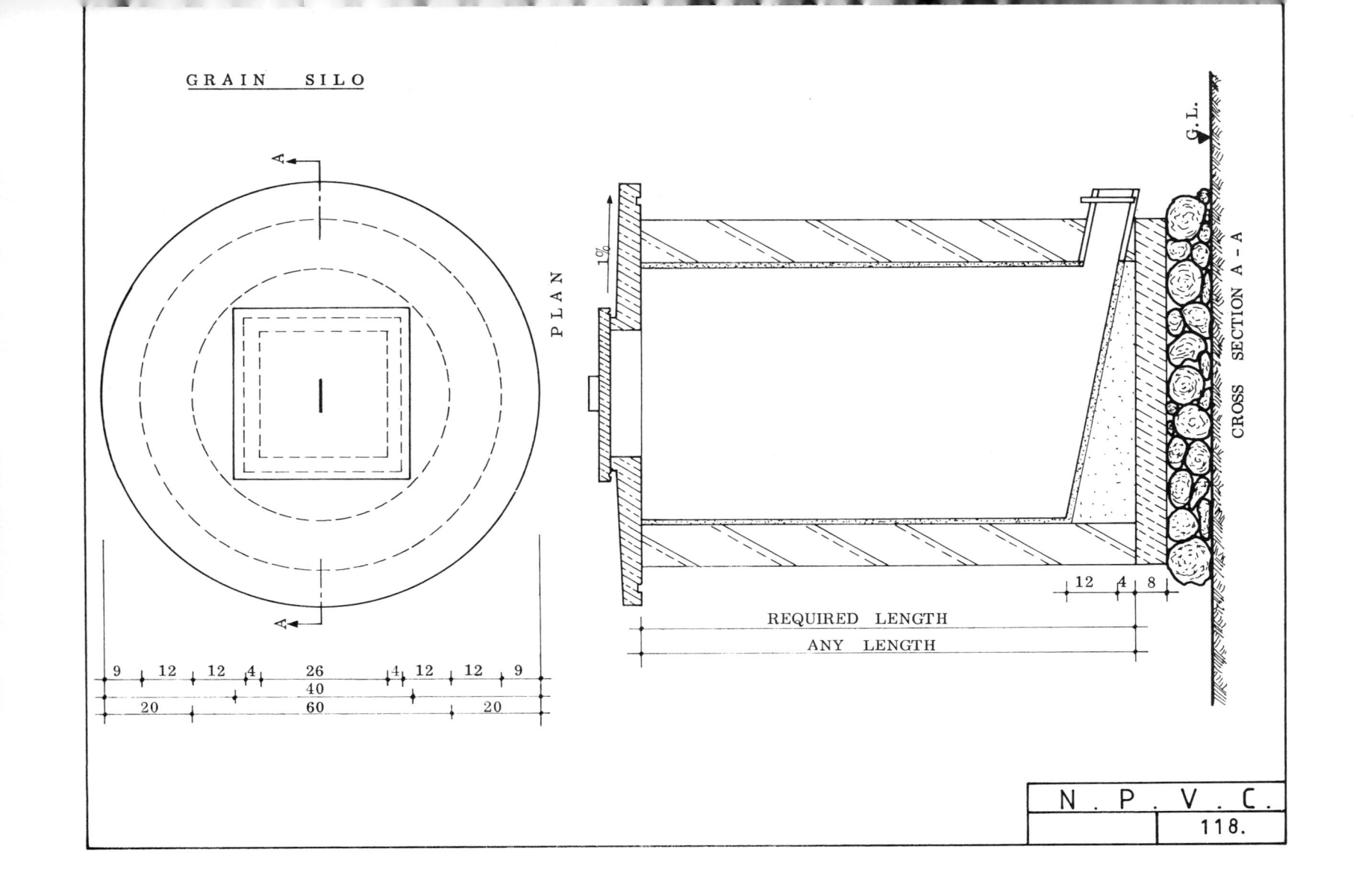

CROSS SECTION A-A

REQUIRED LENGTH
ANY LENGTH

G.L.

PLAN

N.P.V.C. 118.

DRAINAGE

PLAN

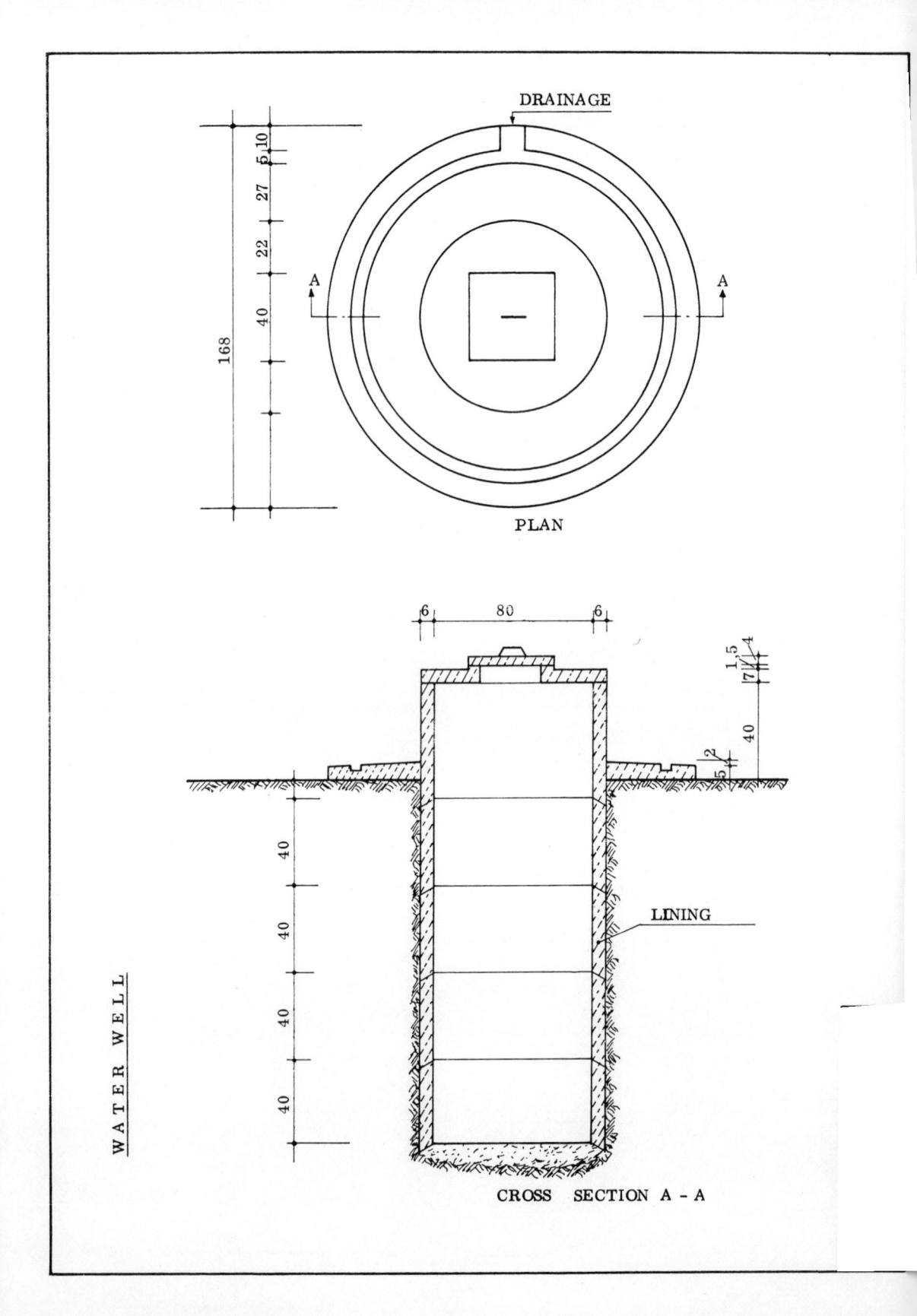

CROSS SECTION A - A

WATER WELL

LINING

WATER FILTER

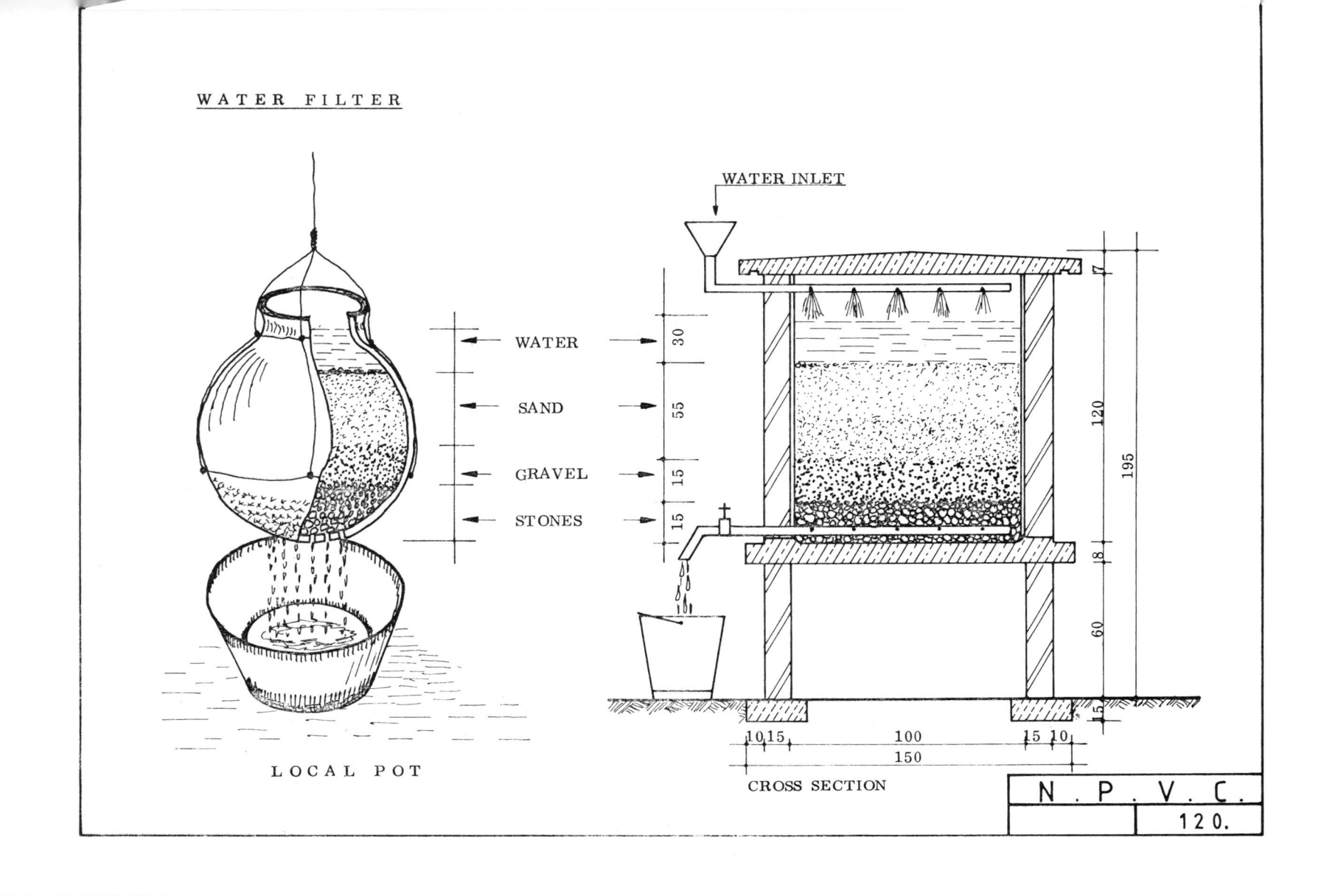

WATER
SAND
GRAVEL
STONES

LOCAL POT

WATER INLET

CROSS SECTION

195
120
60
15
17
8
15
15 10
100
150
10 15
30
55
15
15

N. P. V. C.
120.

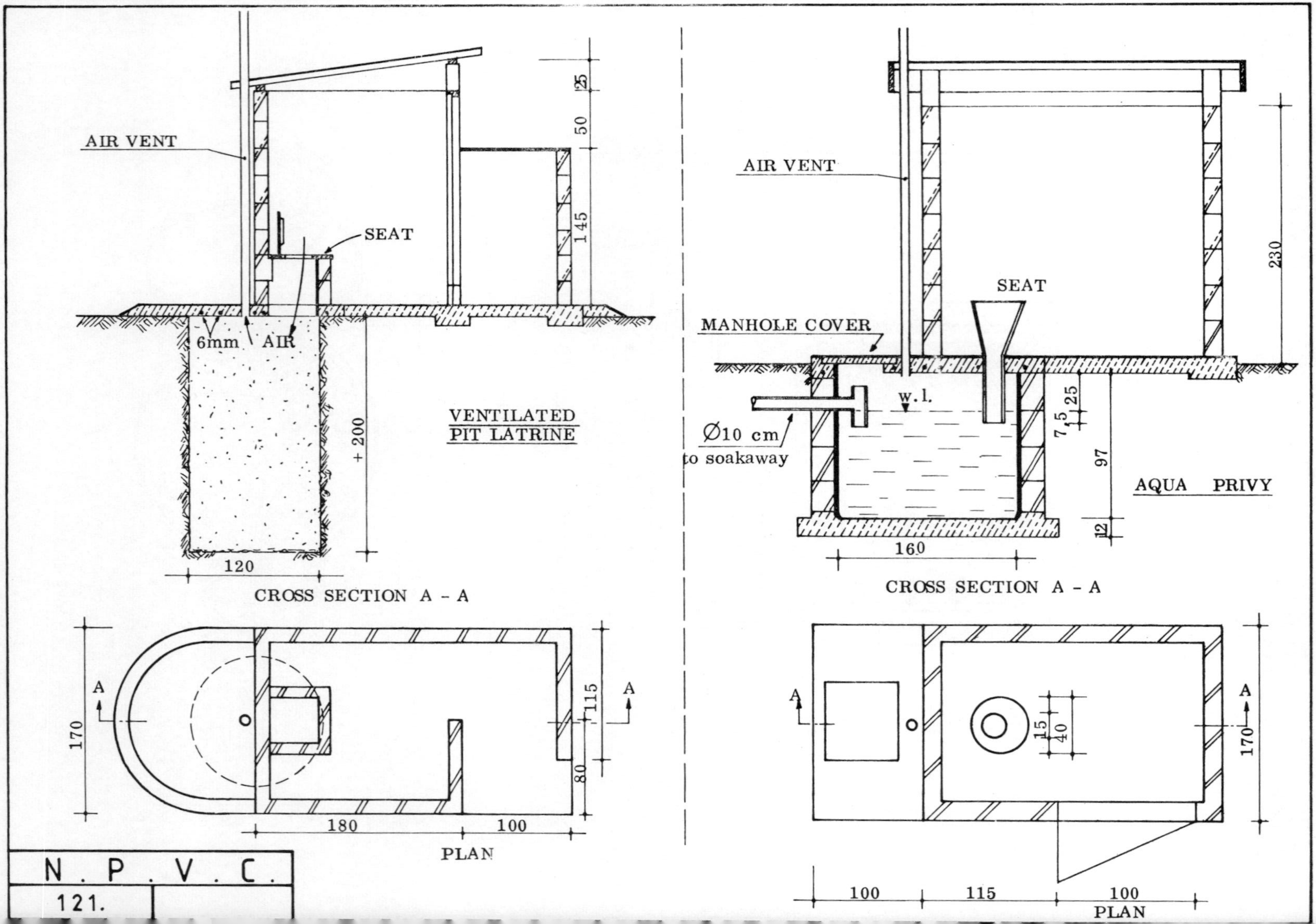

N. P. V. C. 121.

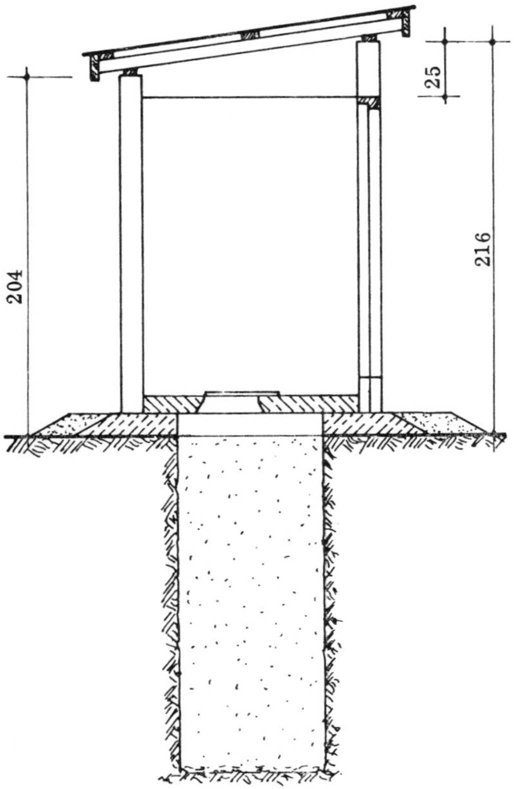

PIT LATRINE

CROSS SEC A-A

PLAN

SQUATTING SLAB

REMOVABLE LIGHT-WEIGHT CONSTRUCTION AND SLAB CAN BE REUSED.

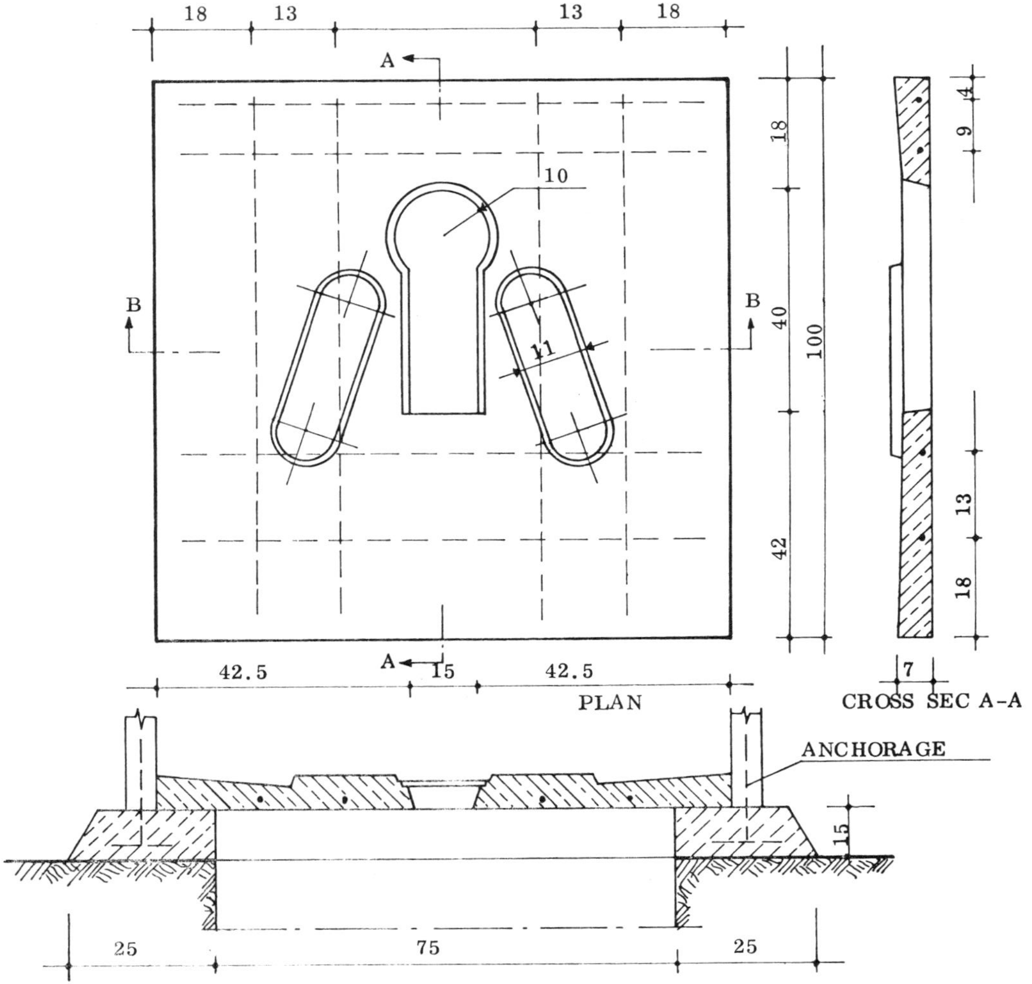

ANCHORAGE

N.P.V.C.

122.

BUCKET LATRINE

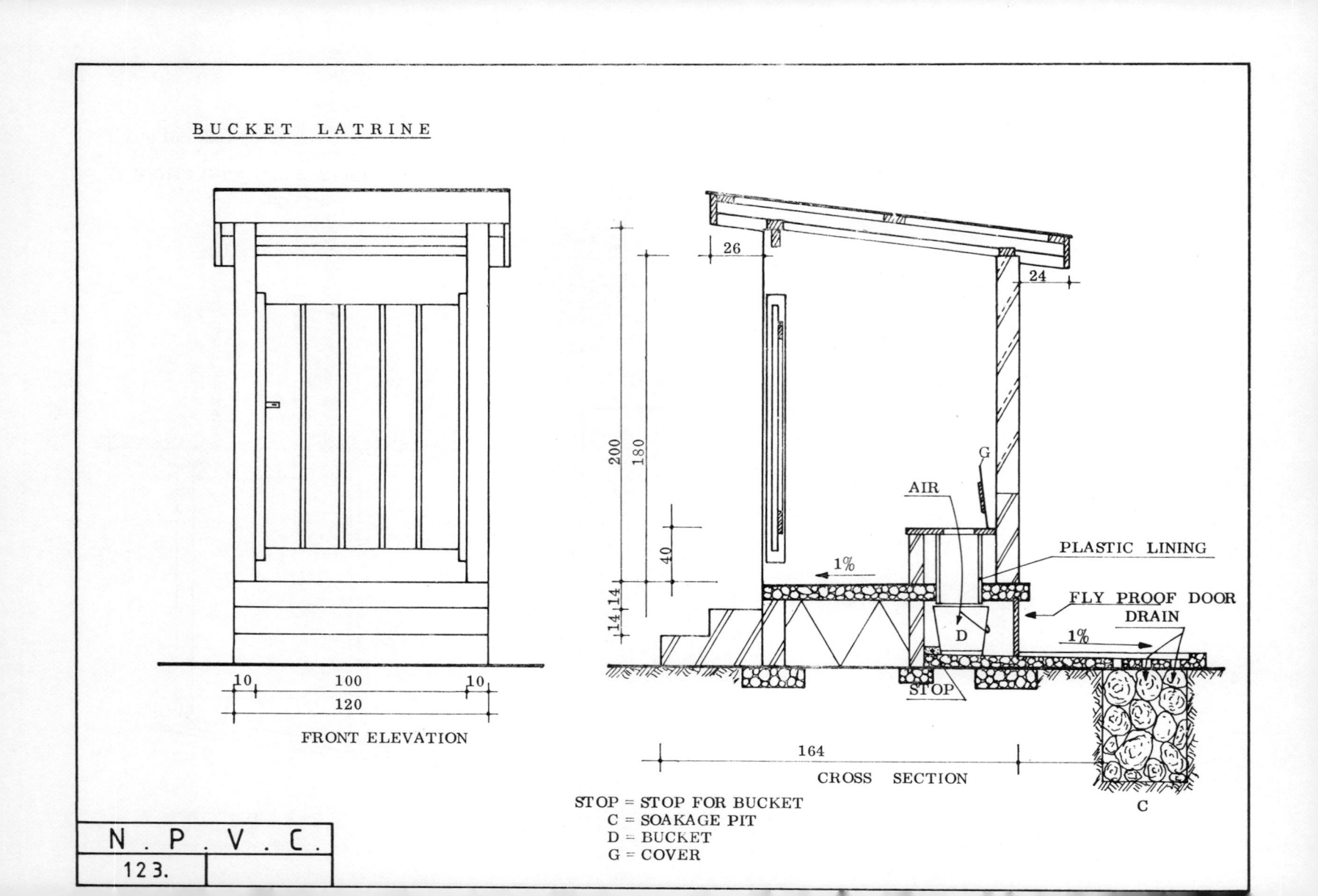

CROSS SECTION

PLASTIC LINING
FLY PROOF DOOR
DRAIN
AIR
G
D
STOP

STOP = STOP FOR BUCKET
C = SOAKAGE PIT
D = BUCKET
G = COVER

FRONT ELEVATION

N.P.V.C.

123.

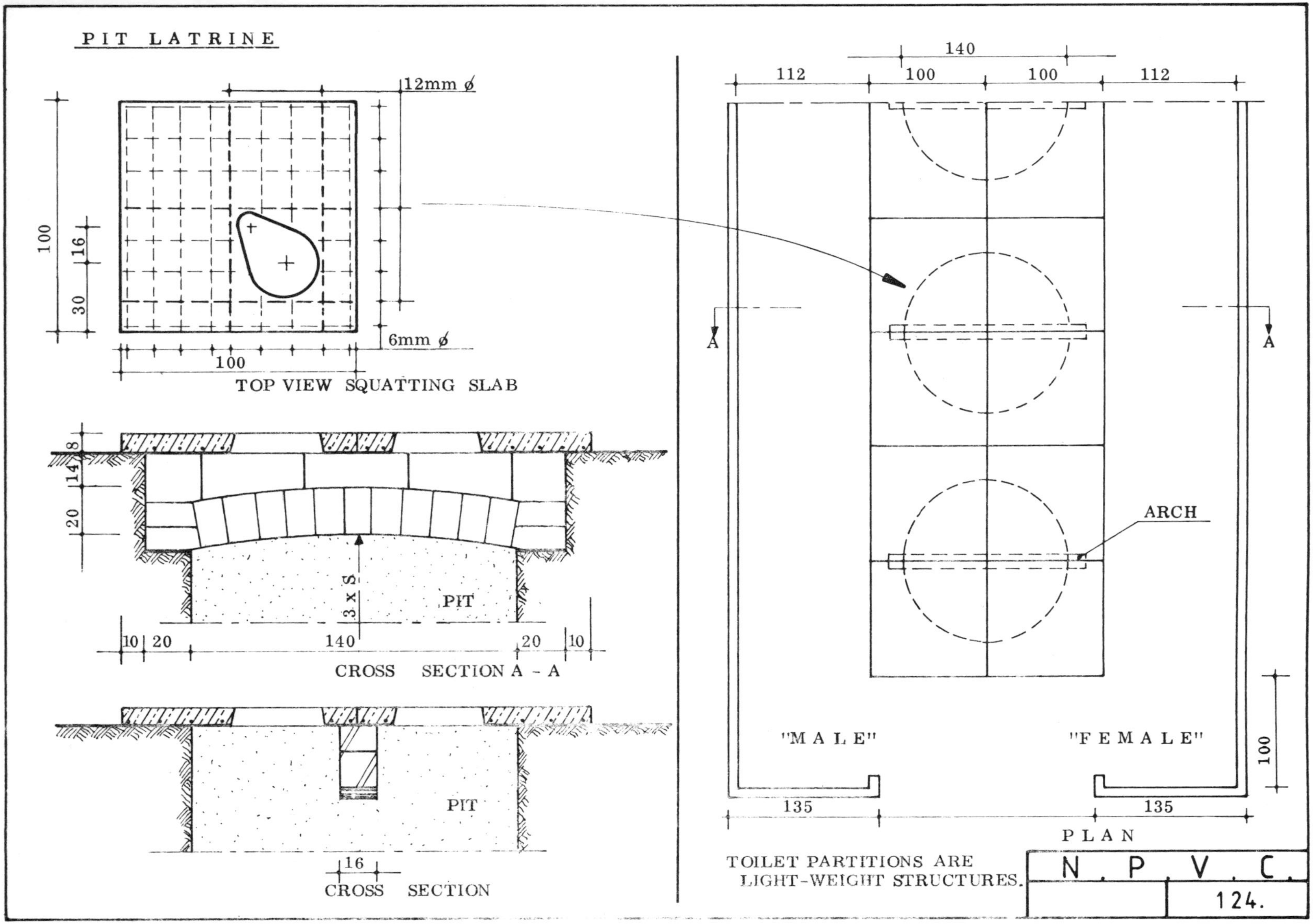

PIT LATRINE

PLAN

"FEMALE"

"MALE"

ARCH

TOILET PARTITIONS ARE LIGHT-WEIGHT STRUCTURES.

TOP VIEW SQUATTING SLAB

CROSS SECTION A - A

CROSS SECTION

PIT

N. P. V. C.

124.

MANHOLE

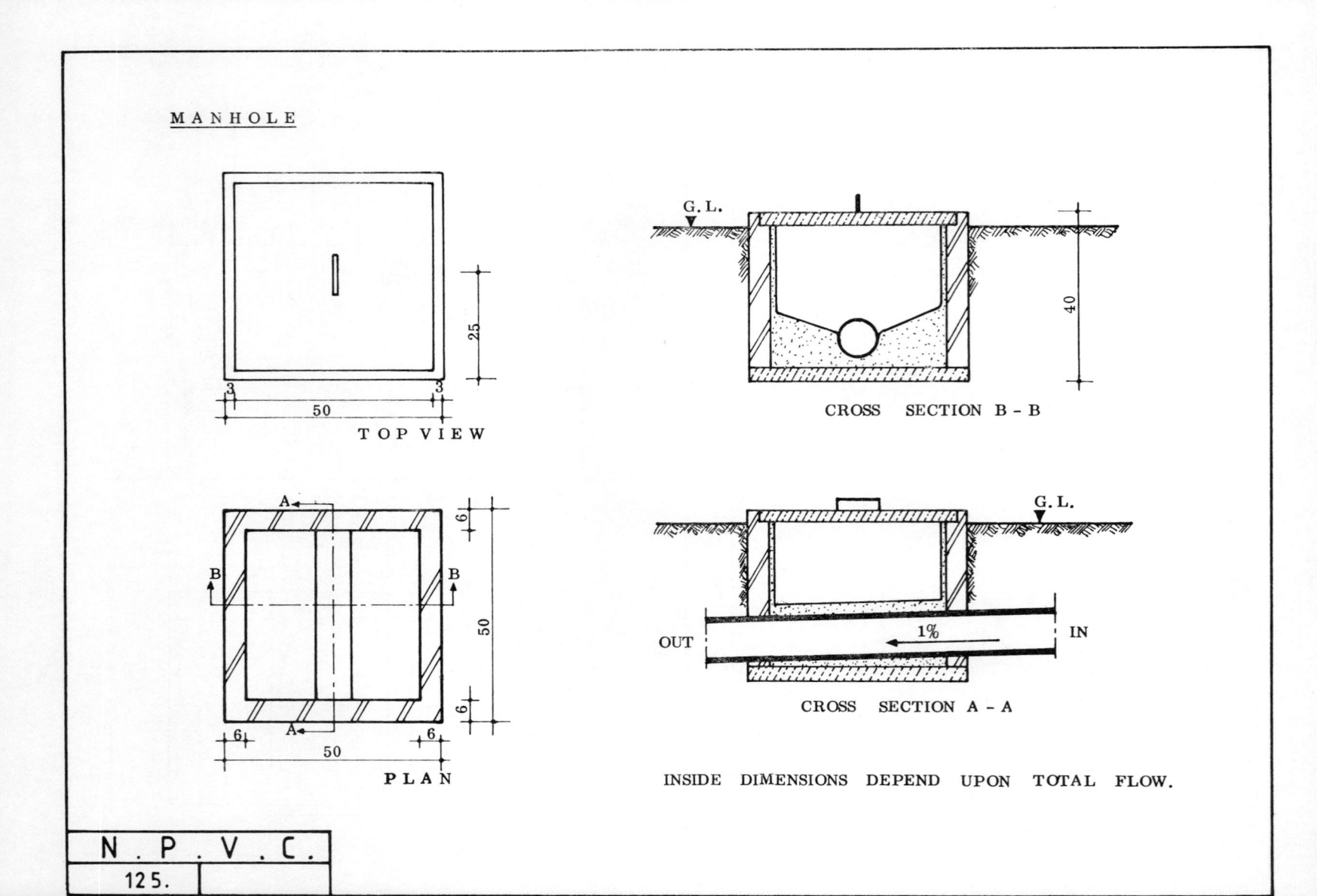

TOP VIEW

PLAN

CROSS SECTION B - B

CROSS SECTION A - A

IN
OUT
1%
G.L.

INSIDE DIMENSIONS DEPEND UPON TOTAL FLOW.

N.P.V.C.
125.

SEPTIC TANK

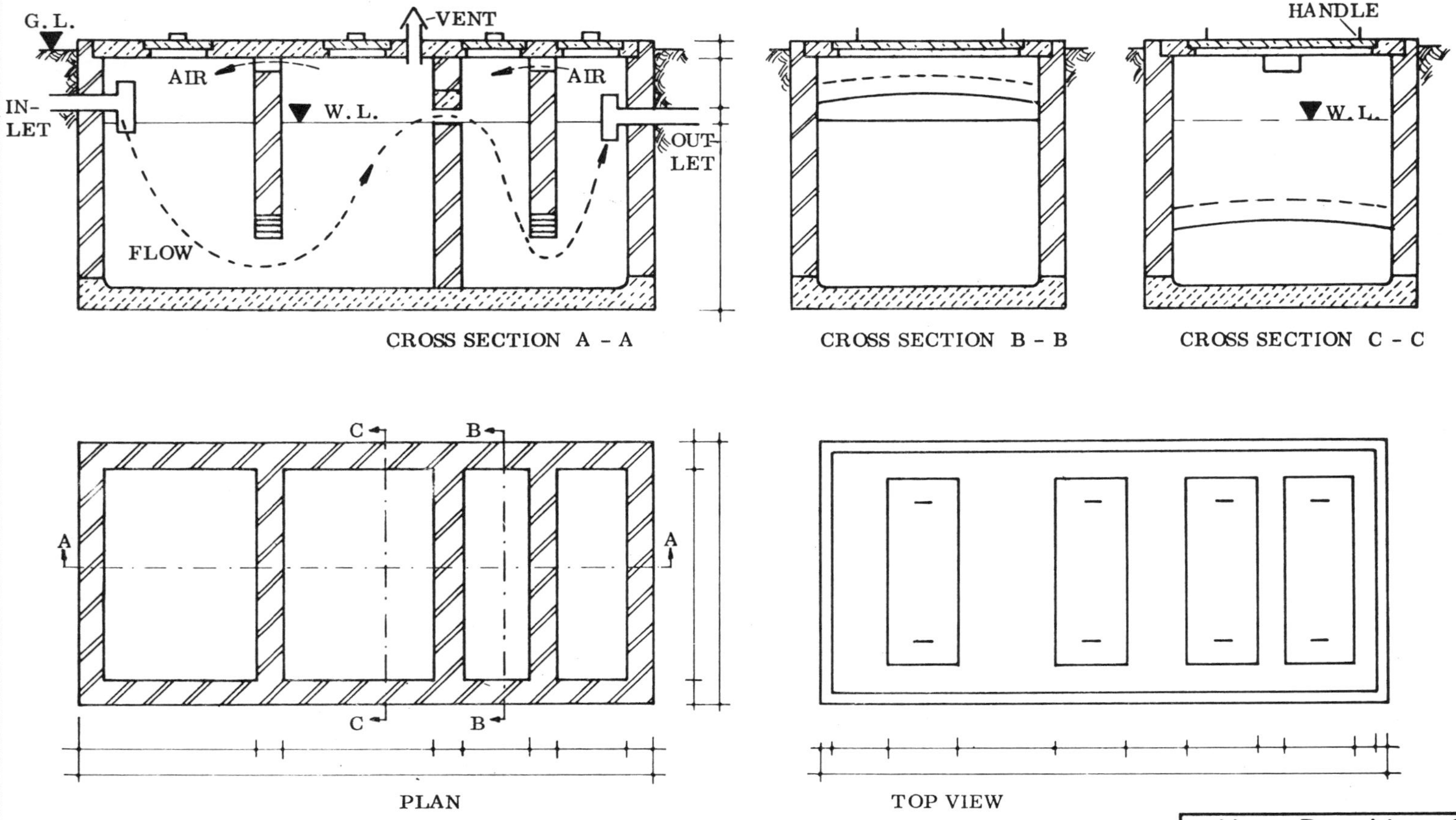

CROSS SECTION A - A

- G.L.
- IN-LET
- VENT
- AIR
- W.L.
- FLOW
- AIR
- OUT-LET

CROSS SECTION B - B

CROSS SECTION C - C

- HANDLE
- W.L.

PLAN

TOP VIEW

DIMENSIONS DEPEND ON THE TOTAL SEWAGE INPUT

N.P.V.C.

126.

NPVC